Action Modular Origami

to intrigue and delight

How to fold paper into geometrical sculptures that move, spin and change shape

Tung Ken Lam

Tarquin

Dedication

For everyone who wants to know how moving folding can be.

ISBN (book): 9781911093947
ISBN (ebook): 9781911093954

Published by Tarquin
Suite 74, 17 Holywell Hill
St Albans AL1 1DT, UK
www.tarquingroup.com

Tung Ken Lam www.foldworks.net

Contents

Introduction **ix**
*Note on Difficulty levels: * grades folding, * grades assembly* ix

Symbols and Procedures **xv**

1 Sliders **1**
1.1 Magic Star: spiral version * * 2
1.2 4-piece Harlequin Slider *** ** 4
1.3 Shaky Shuriken Slider *** *** 6
1.4 Square Slider *** ** 8
1.5 Hexagonal Slider ** ** 10
1.6 Ribbon Slider *** *** 12

2 Flexagons and Rotating Rings **15**
2.1 3-piece Trihexaflexagon ** *** 16
2.2 Rotating Octagram Ring ** *** 20
2.3 12-piece Carousel *** *** 22
2.4 Multiform Flipper ** *** 24
2.5 Double Skew Tetrahedra **** *** 26

3 Magic Wallet Series **29**
3.1 Magic Wallet ** ** 30
3.2 Jacob's Ladder ** **** 33
3.3 Four Square Flexagon ** ** 35
3.4 Four Cube Flexagon ** *** 36
3.5 Six Cube Flexagon (Pseudo-Fluxicube) ** *** 38
3.6 Fluxicube: Rotating Ring of Six Cubes ** *** 39
3.7 Wobbling Wall ** *** 40

4 Spinners and Wheels **41**
4.1 3-piece Square Spinner ** ** 42
4.2 Skeletal Octahedron * ** 44
4.3 WXYZ *** *** 46
4.4 Beak Unit *** *** 48
4.5 Octagon Waterwheel * *** 50
4.6 3D Magic Star ** ** 52
4.7 Octa Slider 3D Wheel *** *** 54

5 3D Shapeshifters **55**
5.1 Radioactive Ball * * 56
5.2 Jitterbug *** *** 58
5.3 Petrie-Coxeter Honeycomb *** **** 61
5.4 Collapsoid *** **** 66
5.5 Flexicuboctahedron *** **** 68
5.6 Skeletal Cube-Rhombic Dodecahedron ***** ***** 71
5.7 Cube-Regular Dodecahedron ***** ***** 73

Further Reading **77**

Bibliography **78**

Index **80**

Notes for Teachers available online

Sliders

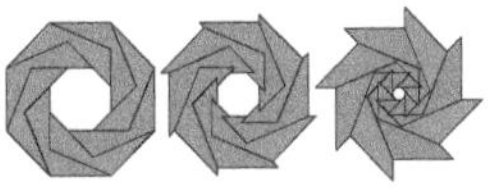

Magic Star: spiral version
** 2

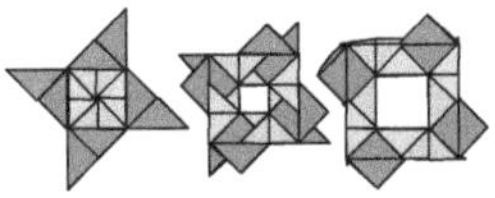

4-piece Harlequin Slider
*** ** 4

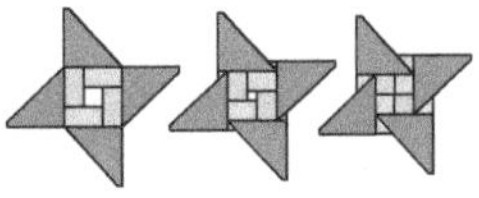

Shaky Shuriken Slider
*** *** 6

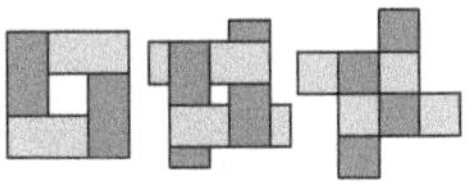

Square Slider
*** ** 8

Hexagonal Slider
** ** 10

Ribbon Slider
*** *** 12

Flexagons and Rotating Rings

3-piece Trihexaflexagon
** *** 16

Rotating Octagram Ring
** *** 20

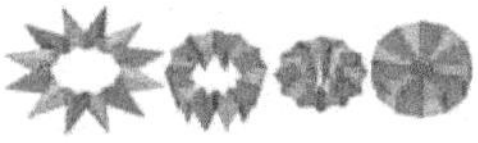

12-piece Carousel
*** *** 22

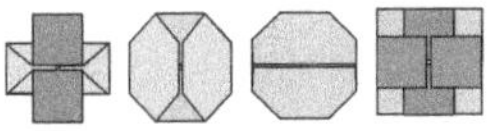

Multiform Flipper
** *** 24

Double Skew Tetrahedra
**** *** 26

Magic Wallet Series

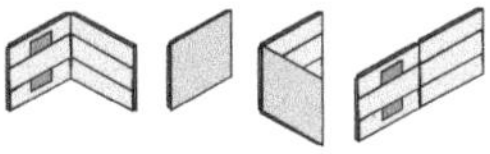

Magic Wallet
** ** 30

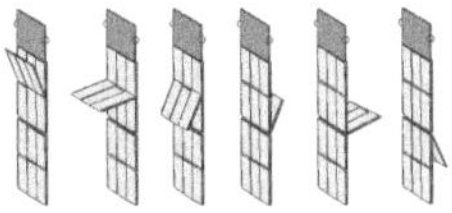

Jacob's Ladder
** **** 33

Four Square Flexagon
** ** 35

Four Cube Flexagon
** *** 36

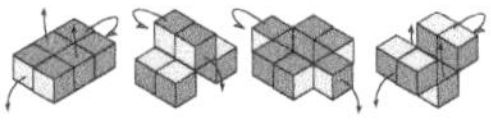

Six Cube Flexagon (Pseudo-Fluxicube)
** *** 38

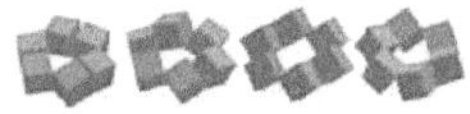

Fluxicube: Rotating Ring of Six Cubes
** *** 39

Wobbling Wall
** *** 40

Spinners and Wheels

3-piece Square Spinner
** ** 42

Skeletal Octahedron
* ** 44

WXYZ
*** *** 46

Beak Unit
*** *** 48

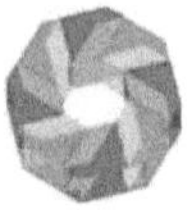

Octagon Waterwheel
* *** 50

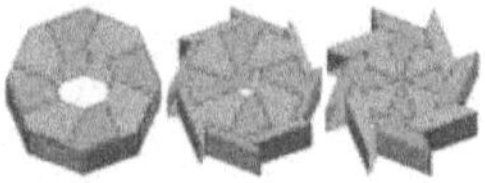

3D Magic Star
** ** 52

Octa Slider 3D Wheel
*** *** 54

3D Shapeshifters

Radioactive Ball
* * 56

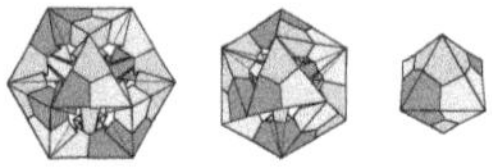

Jitterbug
*** *** 58

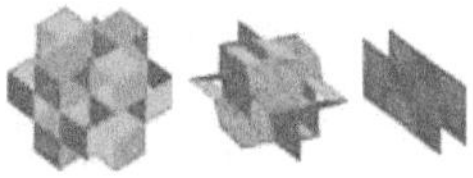

Petrie-Coxeter Honeycomb
*** **** 61

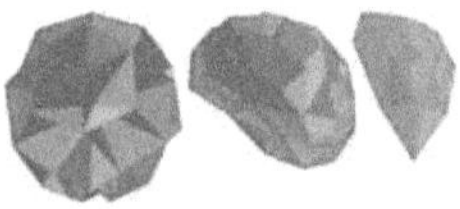

Collapsoid
*** **** 66

Flexicuboctahedron
*** **** 68

Skeletal Cube-Rhombic Dodecahedron
***** ***** 71

Cube-Regular Dodecahedron
***** ***** 73

Introduction

This book is a collection of geometric origami made using the modular technique. You fold a number of simple units that you join together — without glue — like a puzzle. These models all have a bonus feature — they move, spin or change shape in unique ways!

The Models

The projects are grouped in five chapters by how they work:

Sliders are flat models made to expand and contract, forming different shapes.

Flexagons and Rotating Rings are models that fold to expose previously hidden faces. *Rotating Rings* can be turned inside out in special ways.

Magic Wallet Series models use the two-way hinge from a famous magic trick to make flexagons, rotating rings and other action models.

Spinners and Wheels are models that you blow to spin, or turn on a flat surface.

3D Shapeshifters smoothly transform from one 3D shape to another.

Difficulty

Separate difficulty ratings are given for folding the units and then for assembly.

Difficulty of folding the units

Simple * Valley and mountain folds only, no more than about a dozen steps e.g. *Radioactive Ball* and *Skeletal Octahedron*.

Low Intermediate ** Only a few compound folds like reverse folds and squash folds e.g. *Three-piece Tri-hexaflexagon*.

Intermediate *** Several compound folds like reverse folds and squash folds e.g. *WXYZ*.

High Intermediate **** May have a few tricky folds like multiple sinks or need careful adjustment for well-fitting units e.g. *Double Skew Tetrahedra*.

Complex ***** High accuracy is essential. May have several difficult folds or need extensive precreasing e.g. *Cube-Regular Dodecahedron*.

Difficulty of assembly

Simple * Each unit attaches easily to the next and does not come apart easily e.g. *Radioactive Ball* and *Magic Star: spiral version*.

Low Intermediate ** Each unit attaches easily to the next but is only partially stable *Skeletal Octahedron*.

Intermediate *** Each unit attaches fairly easily to the next but is not particularly stable until several units are combined e.g. *Multiform Flipper*, or attaching each unit is not straightforward e.g. *Octagon Waterwheel*, or the last unit is difficult to attach e.g. *WXYZ* and *Three-piece Tri-hexaflexagon*.

High Intermediate **** Careful attention required for the order of joining e.g. *Petrie-Coxeter Honeycomb* and *Jacob's Ladder*.

Complex ***** High accuracy is essential. Units do not securely attach to the next, or are awkward to join e.g. *Skeletal Cube-Rhombic Dodecahedron*.

Tips on folding and paper

Paper

You should be able to make all of the models in this book using commonly available paper. Ordinary 80 gsm A4 photocopy paper is readily available in a variety of colours. For squares you can cut your own squares from rectangles, use paper from memo cubes or 15 cm origami paper. Origami paper usually has different colours on each side: this can help with following diagrams and improve the appearance of some finished models e.g. *Three-piece Square Spinner*. Other times paper that is the same colour on both sides is easier to obtain and works better for some models e.g. it can prevent the white side of the paper showing in *WXYZ*.

In general, choose papers that are strong (will not rip easily), not too thick (makes thick layers and causes inaccuracy) and which will take and hold a crease well. This means you should normally avoid newsprint, paper-backed foil and card stock. However, if you want to make large models then you might need to use slightly thicker paper.

You can use plain paper for all of the models in this book. Patterned paper can enhance some models, but you need to choose suitable paper. Very busy patterns can be overwhelming and distracting in the finished model. For very fine work, choose acid-free paper that will not discolour with age.

Reusing paper like junk mail and flyers is fine, especially for practice, but be careful if the paper has personal data. You can fold paper from magazines but this can be unpleasant as the coating rubs and cracks off the coated paper.

The *Magic Wallet* series of models uses a special source of recycled paper: train tickets that are thin card stock, 85 by 54 mm. These tickets have an added advantage: tucking in edges with rounded corners is easier than with square corners. If you ask a station employee for used tickets, explain your request is for origami as they might decline to prevent ticket fraud.

Paper size

Convenient paper sizes are suggested for each model so that the results fit your hands. If you want to know the size of the result for other paper sizes, use the finished model ratio given in the diagrams. The ratio is usually the longest length in the largest configuration. For example, the ratio of 1.8 for the 12-piece Carousel means that squares of length 1.0 make a model with a diameter of 1.8. If the starting rectangle is not square, the shortest length is assumed to be 1.

Preparing paper

Most of the models use rectangles of convenient proportions like squares, 1:2, 1:4, silver and bronze rectangles.

Making 1:2 or 1:4 rectangles from a square is straightforward: fold in half or into quarters and cut.

Making a square from a rectangle is only slightly harder: bisect a corner by folding the shorter edge onto the longer edge. You have now found the point on the longer edge that makes a square.

Here is a brief explanation about silver rectangles for those unfamiliar with the term. This term applies to rectangles with the proportion $1{:}\sqrt{2}$ and also to paper from the international standard A series. This standard defines A0 to have an area of 1 m^2 and sides in the ratio $1 : \sqrt{2}$. A1 has half the area of A0 but also has sides in the ratio $1 : \sqrt{2}$. A2 has half the area of A1, and so on. The common A4 paper has sides 210 by 297 mm (8.27 in by 11.7 in). US letter paper is 8.5 in by 11 in (215.9 mm by 279.4 mm) so is about 6 mm (0.24 in) wider and 18 mm (0.71 in) shorter than A4. To convert US letter to a silver rectangle, trim a long strip $\frac{3}{8}$ in (9 mm) wide.

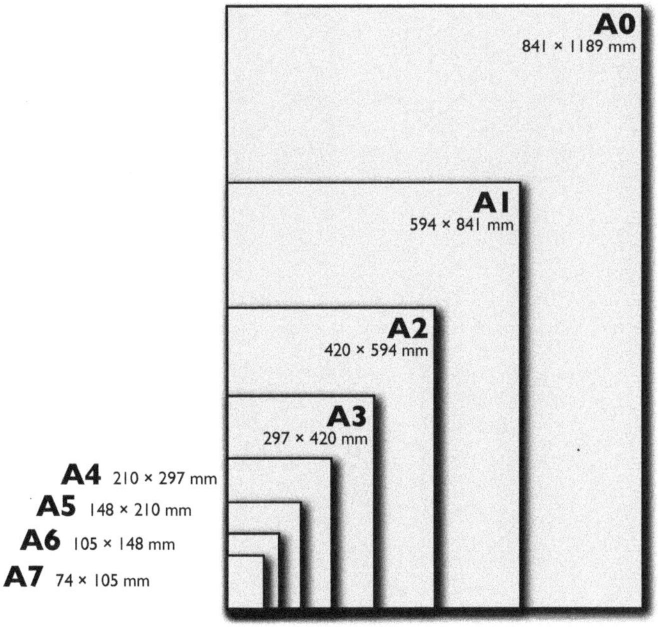

Making a square from a rectangle

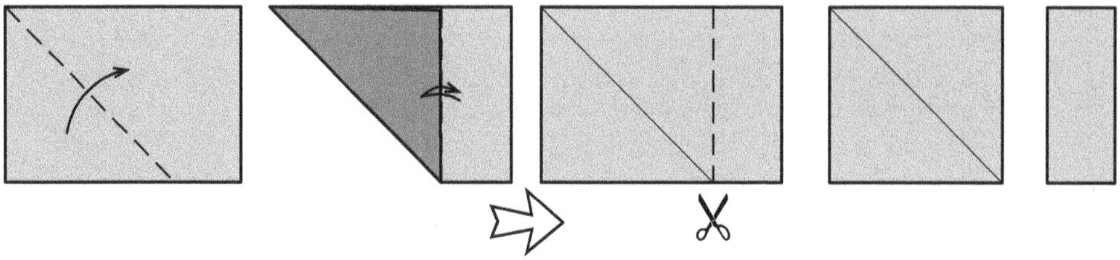

Making a silver rectangle from a long rectangle

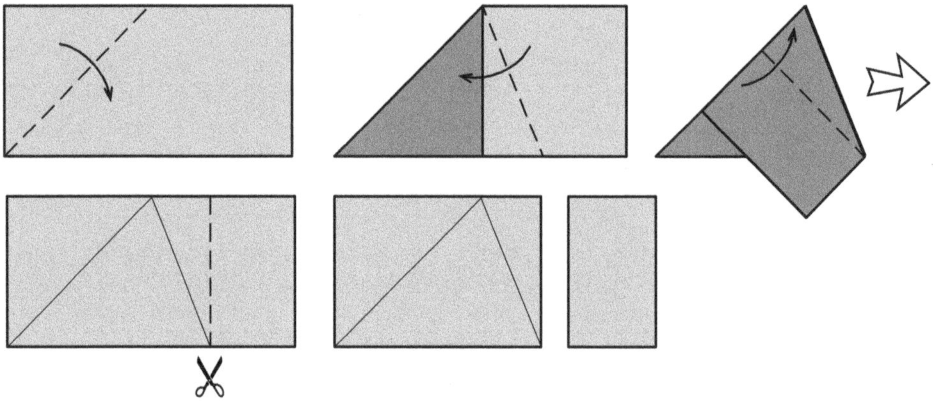

Making a silver rectangle from a square

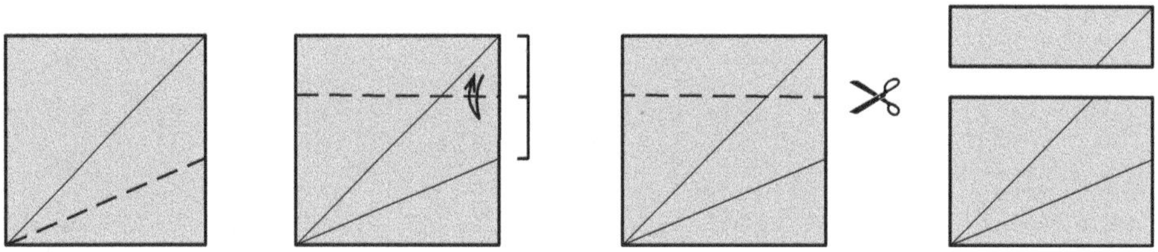

Making a silver rectangle from a short rectangle

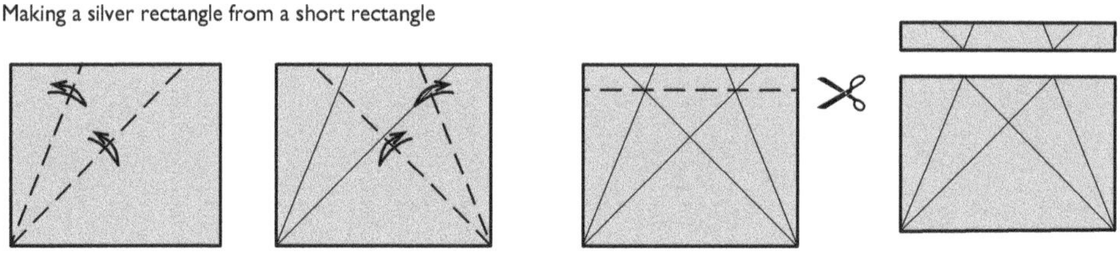

Making silver rectangles from non-silver rectangles is fairly straightforward. The diagonal of a unit square is $\sqrt{2}$, so it's easy to find the point that defines a silver rectangle. In practice there are three methods: one for a rectangle longer than a silver rectangle, one for a rectangle shorter than a silver rectangle and one for a square.

For cutting, certain kinds of envelope openers are convenient to use. They are small, portable, have no exposed blade and are relatively easy to use. You fold the paper where you want to cut and then slide the envelope opener along the folded edge. However, they work best for cuts perpendicular to a raw edge and are not so good for other angles e.g. 60°. Remember not to crease too hard or the blade will not catch properly.

Folding the modules

You can fold paper almost anywhere, but if you have a choice then use a bright and comfortable place for folding. Make sure you have enough suitable paper and have clean hands before you begin. Paper absorbs moisture and oil from your fingers so try to handle the paper as little as possible.

Like most modular origami, sharp accurate folding will make assembly easier. Fold on a hard surface and be careful to keep edges or location marks lined up as you fold.

As you fold more units, you should find your accuracy improve as your "muscle memory" learns the steps and you may find better ways to make the units than the diagrams suggest. This is why you should normally avoid an "assembly line" method of folding such as step 1 on each sheet of paper, then step 2 on each sheet, then step 3 on each sheet, etc. Not only is this dull and unsatisfying compared with making complete modules one at a time, you might find that you've made the same mistake on all of the modules!

Assembling the modules

Once you have two modules try joining them together as this makes folding the modules more motivating. It also lets you find out if you need to make any minor adjustments to the modules, perhaps because of the size and thickness of your paper. This brings me to another tip on assembly: usually you are better to keep adding modules to the same model. Do not make separate sub-assemblies and then join them together — this can become confusing and is usually inefficient.

2D models are mostly straightforward to assemble. You usually need to gently bend the model to join the last module. 3D models are usually harder to assemble than 2D models. If the last module is hard to add, loosen the other modules and try again.

For difficult and awkward assemblies, you might need some tools. The first type of tool lets you poke and tuck flaps in difficult places — cocktail sticks and nail files work well. The second type of tool temporarily holds modules together — you can use paperclips, mini clothes pegs or pressure-sensitive masking tape depending on the model. Paperclips can leave marks so do not leave in place for too long. Also do not leave masking tape on for too long either as it might bond to the paper and rip when removed.

Following the instructions

Diagrams are like a comic strip that you follow, performing the folds in each step. Do not let the origami terminology intimidate you: you can usually tell what's required by looking ahead to the next step. If you've played guitar, diagrams are more like tablature than standard notation as they show you exactly you what to do — there's no need to interpret the notation wondering which string and fret you should play *C* on. However, you can turn the paper to make it easier to fold — the paper does not have to be in the same orientation as the diagram. For example, to make a mountain fold you might find it easier to turn the paper over, valley fold and then turn over again. Or a vertical valley fold is easier to make if your rotate the paper 90°so that you are folding away from your body.

If you are a beginner then start with the simpler models first. Models earlier in the book have more detailed instructions. Later models tend to be briefer, not only because it saves space, but also because as you become more experienced, you can work with larger "chunks" of information. In general the simpler models are at the beginning of each group of models. Also note that some models have different difficulties of folding the modules and assembling them. For example, the modules of *Octagon Waterwheel* are very simple, but are not so simple to assemble.

Beginners usually appreciate the detailed text in each step, especially for assembling the modules. Many experienced folders skip the words in each step: besides fluency, one reason is that they often fold from diagrams written in languages that they cannot read, so manage without reading the words. Even so, the words can still help you by using your verbal communication channel and working memory. Sometimes the words give helpful tips that you might miss if you do not read the diagram carefully, for example only fold between the existing crease lines.

Note that diagrams are usually drawn with the layers of paper slightly spread out, even if they are meant to align exactly. This helps you understand how the layers are arranged. If you are meant to leave gaps, this will be specifically highlighted. As mentioned elsewhere, diagrams use different colours to distinguish the two sides of the paper. However, for most models you can use paper that is the same colour on both sides: look at the final model to decide which kind of paper you should use.

If you are stuck, study the diagram carefully: is there a line or arrow that you've missed? Do the flaps and layers of paper correspond to the paper in your hand? If taking a break and trying again doesn't work, ask someone else for help. It doesn't have to be an origami expert as a fresh pair of eyes might be all that's needed. If nobody is able to help, try starting again from the beginning, carefully following every symbol and instruction. Despite the best intentions of all involved, diagrams may have mistakes so see if you can work around the problem: can you carry on even though you've missed out a step?

See Symbols and Procedures for more hints on understanding diagrams.

Symbols and Procedures

The diagrams in this book use symbols based on standards developed by Akira Yoshizawa, Robert Harbin, Samuel Randlett and others. If you spend some time to understand these symbols then they will soon become second nature. Not only will this save you time and make following diagrams more pleasant, you will also be able to fold from diagrams in languages that you cannot read. You will also develop the valuable skill of folding imaginary paper in your mind.

At each step you can usually look ahead to see the result that you are aiming for. However, sometimes you will need to look carefully at the symbols. Most often you will need to distinguish valley (concave) and mountain (convex) folds: the former uses a dashed line and a plain arrowhead; the latter uses a line of dashes and dots with a hollow half arrowhead. See overleaf.

The standard symbols have some quirks. The turn over symbol might evoke turning over the model, but taken literally would turn the model over again. The cut symbol uses scissors but using a knife is better in practice — the scissors are a familiar design trope for cutting coupons out of magazines, etc.

Some symbols are not always used. As the diagrams in this book use different shading for the two sides of the paper, most diagrams omit the symbols coloured side up / down. Repeat symbols are rarely used as it's usually obvious where you need to repeat the steps — read the text for specific information about repeats. For a similar reason, the enlarged view symbol is not always used; and curiously, no standard reduced view symbols exists.

Some common compound folds are given special names. See diagrams on page xvii. Demonstrations online will be available — indicated after the title by a play button symbol ▶. This book only refers to a few procedures by name: the squash fold, (inside) reverse fold, outside reverse fold, rabbit ear, preliminary fold and waterbomb base. Most models only use valley and mountain folds with a few squash and reverse folds. Rabbit ears are very rare in this book, and they are not those typically used in figurative origami. Typical rabbit ears use the angle bisectors of a triangle, but the rabbit ear from a rectangular flap uses two angle bisectors and one perpendicular bisector. The preliminary fold and waterbomb base are related: turn one inside out and the result is the other.

You will find two other procedures used in some models: dividing a length into equal parts and folding 60°. It is easy to divide a length into halves, quarters, eighths but thirds and fifths can be also made without a ruler. A good way is Shuzo Fujimoto's iterative method. Folding 60°is a special case of Abe's trisection of an acute angle.

Folding lines and symbols

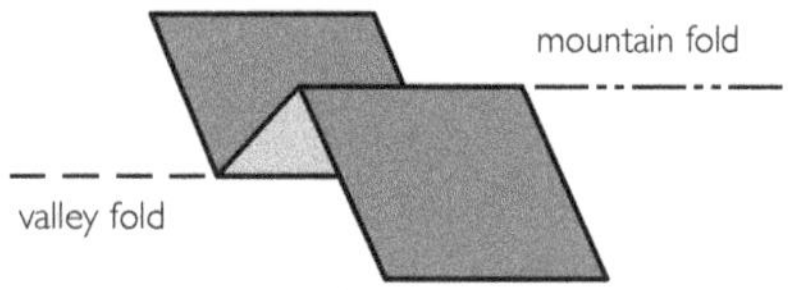

 fold to the front

 fold to the rear

 crease and unfold

 pleat

 fold over and over

 fold placing dot onto circle

 push in / apply pressure

 hold the locations indicated by the circles and perform the action described e.g. pull apart

 open up

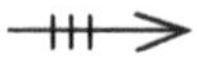 repeat three times (or by the number of strokes)

Other symbols and lines

 start with the coloured side of the paper up / down

raw edge

existing crease

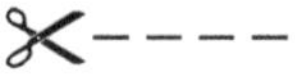 cut

 hidden line, previous/next position or imaginary line

 length divided into equal divisions e.g. thirds

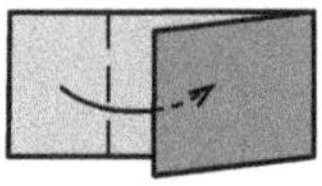 tuck inside

Symbols between diagrams

 turn the model over

 viewpoint for the next diagram

rotate the model by the angle shown in the given direction e.g. 90 degrees clockwise

 next diagram shows an enlarged view

Squash fold ▶

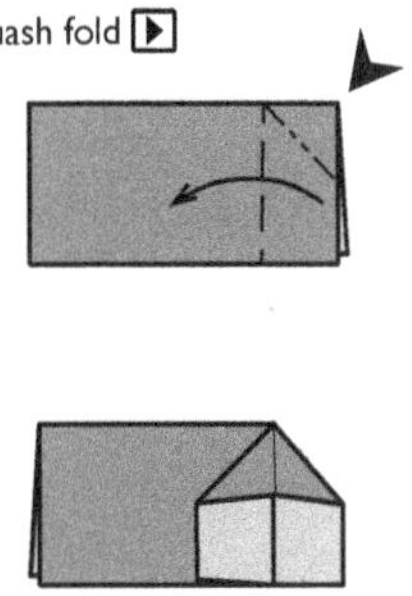

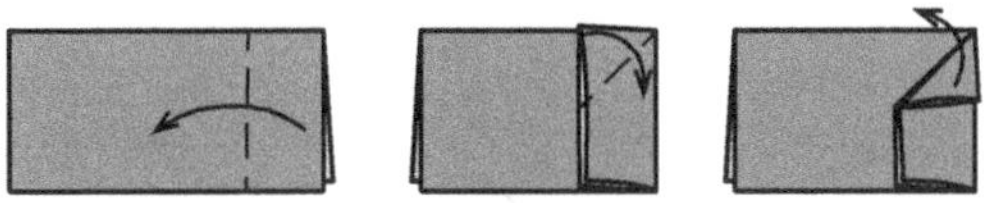

Optional precrease

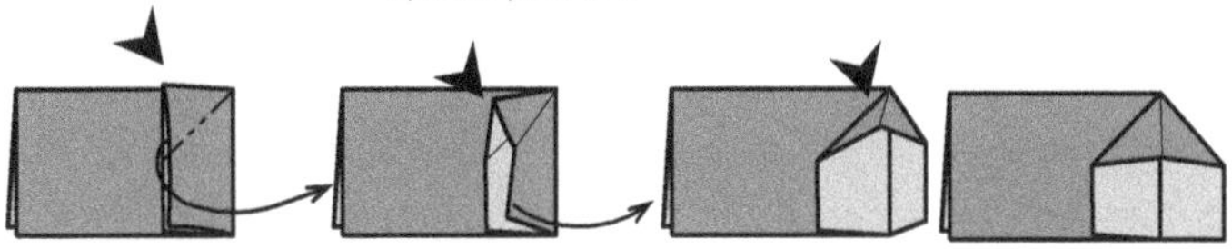

Use a finger to open the pocket. Slide your finger upwards to fully open the pocket. Use your other hand to flatten the pocket.

Inside reverse fold ▶

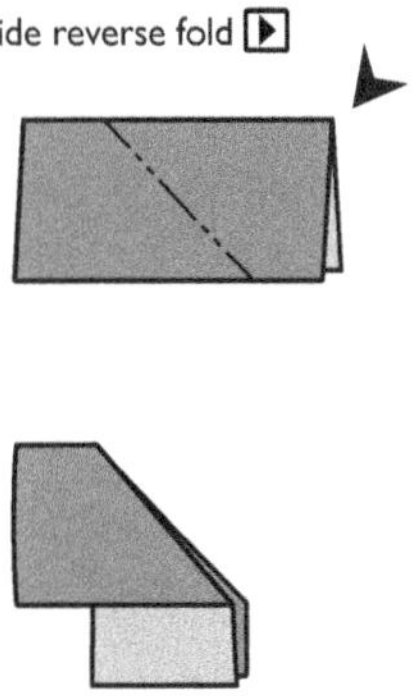

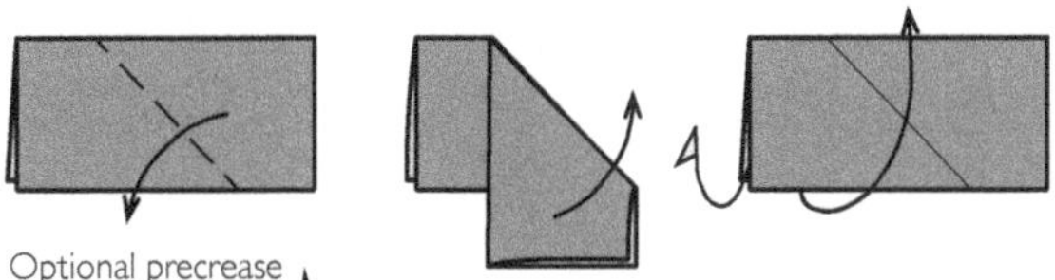

Optional precrease

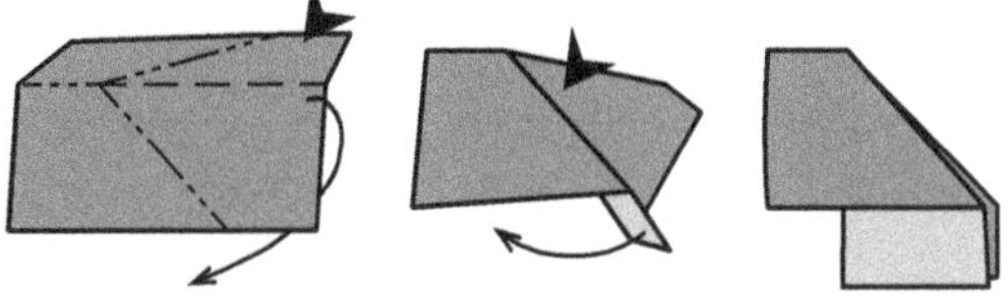

Open the paper and push in the right side. Change the mountain folds to the right and front into valley folds. Flatten.

Outside reverse fold ▶

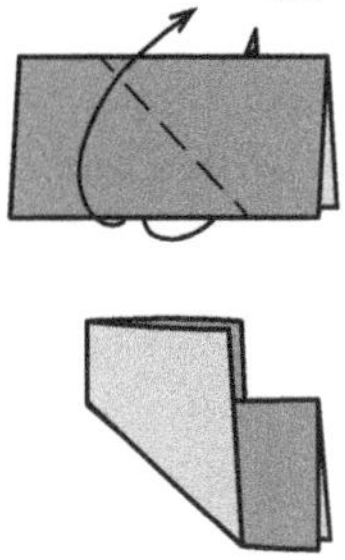

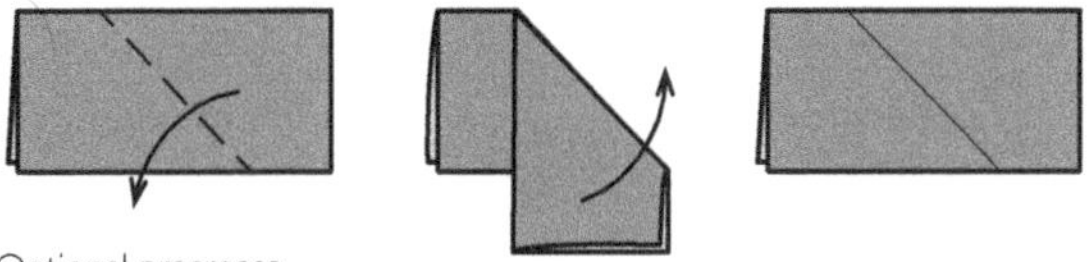

Optional precrease

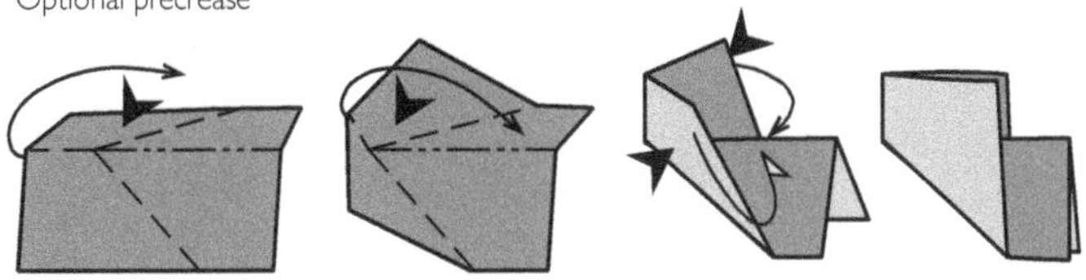

Open the paper and push the point from which creases radiate. Change the mountain folds to the left and rear into valley folds. Flatten.

Rabbit ear ▶

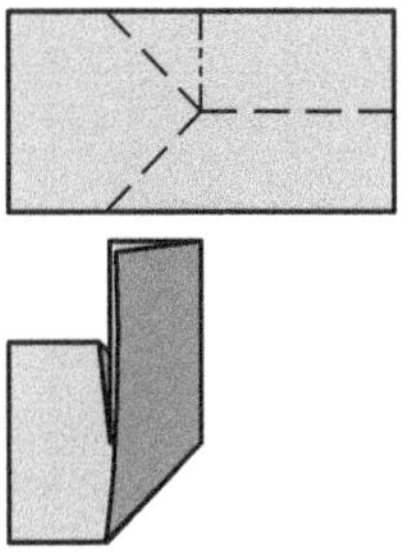

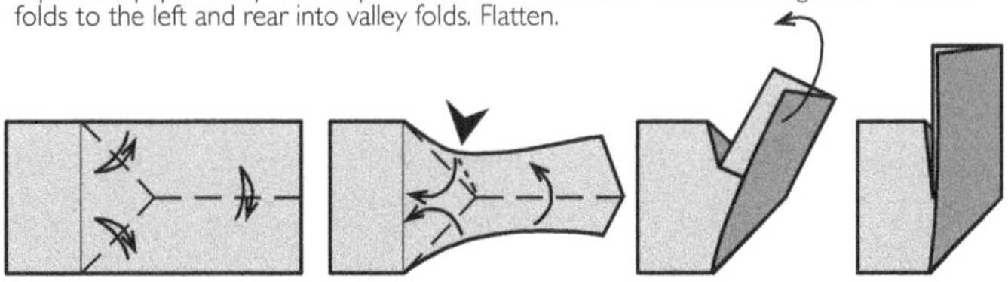

Precrease two angle bisectors and one perpendicular bisector. Pinch together the two long edges to form a flap. Raise this flap and flatten to one side.

Making a waterbomb base

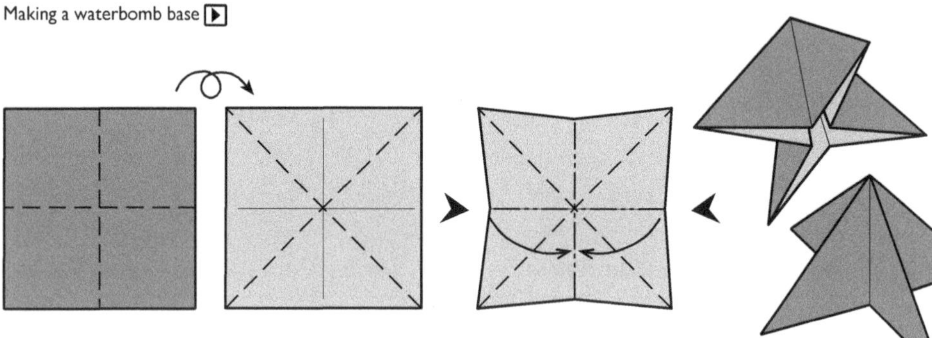

Making a preliminary fold

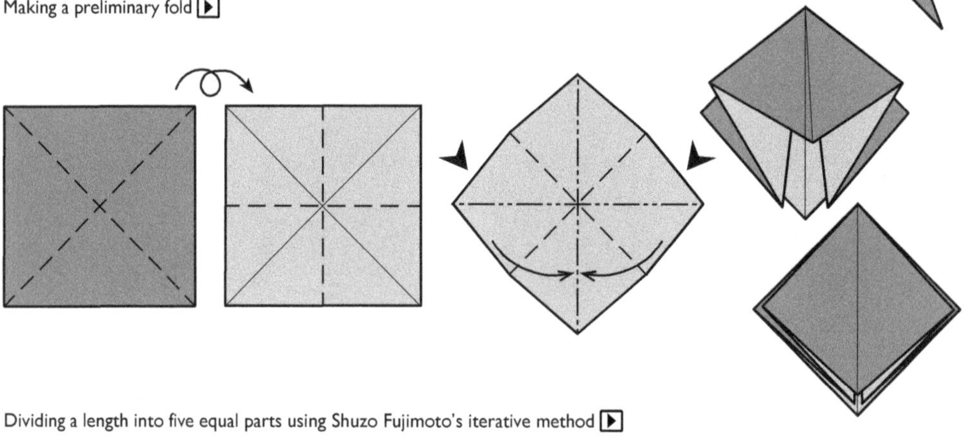

Dividing a length into five equal parts using Shuzo Fujimoto's iterative method

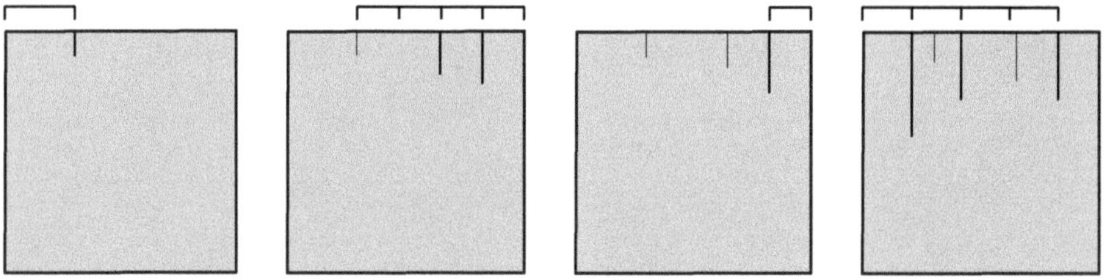

Start by guessing one fifth.

The rest of the length is "four fifths"; divide this in half and divide in half again.

This new "one fifth" is a better approximation to one fifth.

Repeating this process quickly reduces the error to practically nothing.

Trisecting a right angle

Fold the lower left corner onto the middle line. At the same time the fold starts at the top left. This fold appears in different guises but is essentially the same fold.

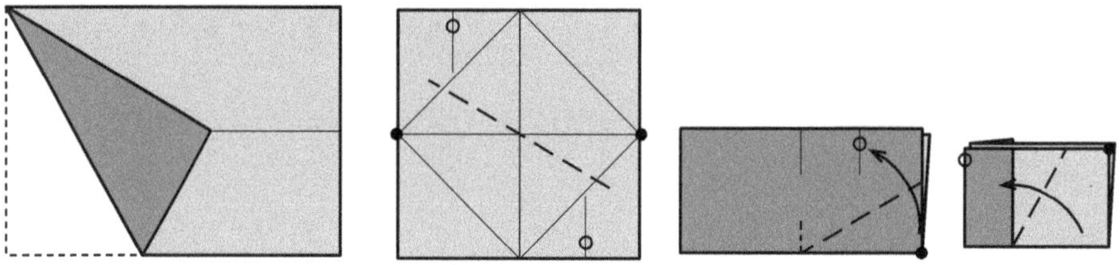

Chapter 1

Sliders

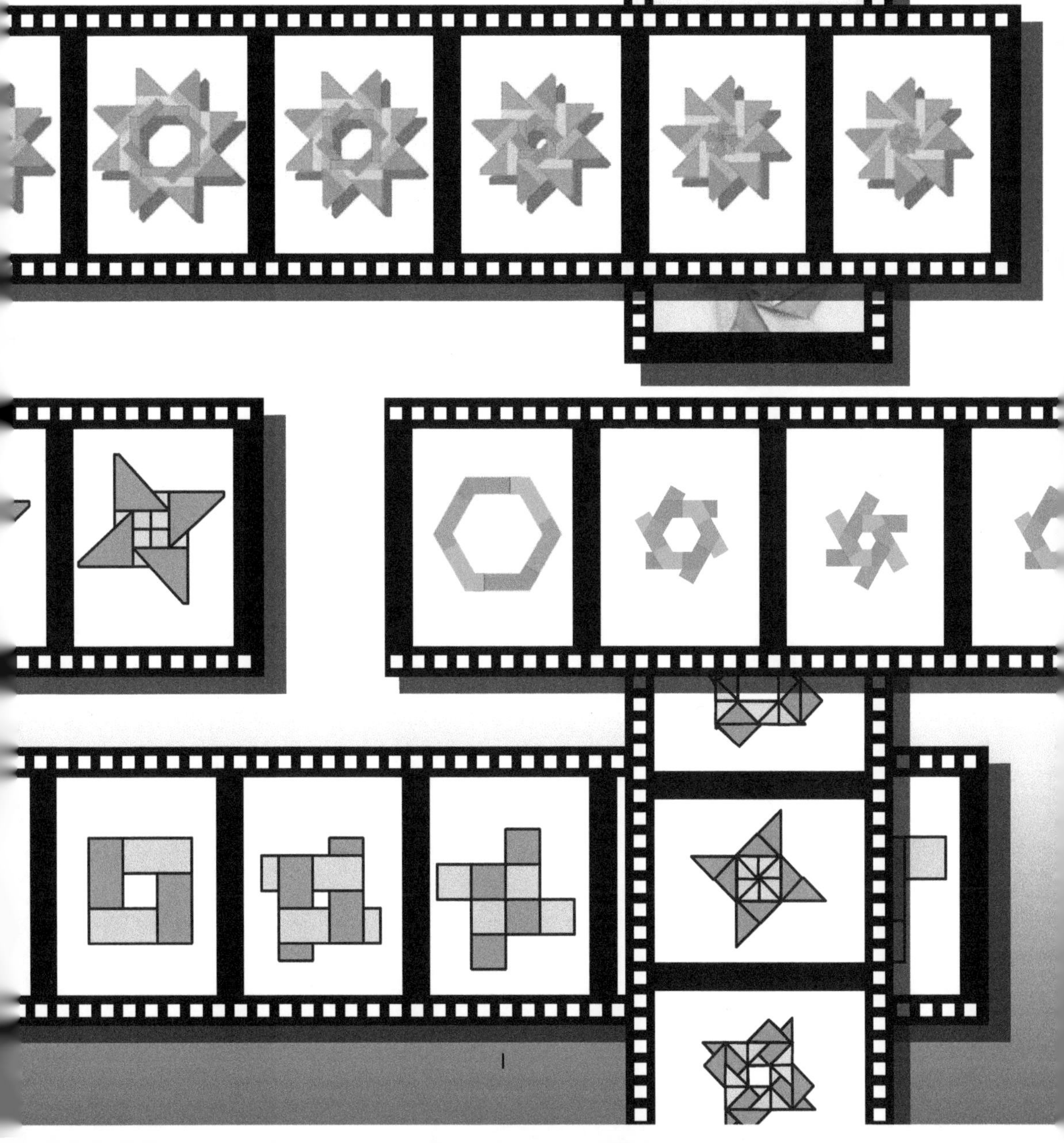

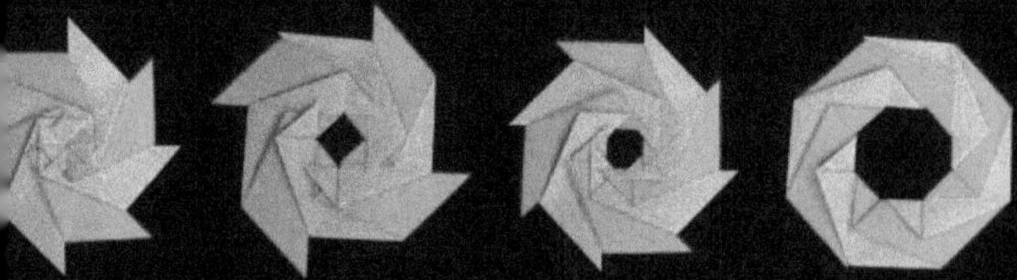

Magic Star
Spiral Version
★☆☆☆☆
Use 8 squares
▶ Finished model ratio 1.4

The result of step 5 is Robert Neale's original unit. His classic *Pinwheel-Ring-Pinwheel* appears in *Origami, Plain and Simple* by Tom Hull, p. 36-37 as well as many other places. By making one extra fold in step 6, the appearance changes dramatically on one side. Neale's units are stable when at least two are joined, but this spiral version is only stable when at least three units are joined. Three other models in this section use the same type of sliding mechanism as Neale's original.

Module ★

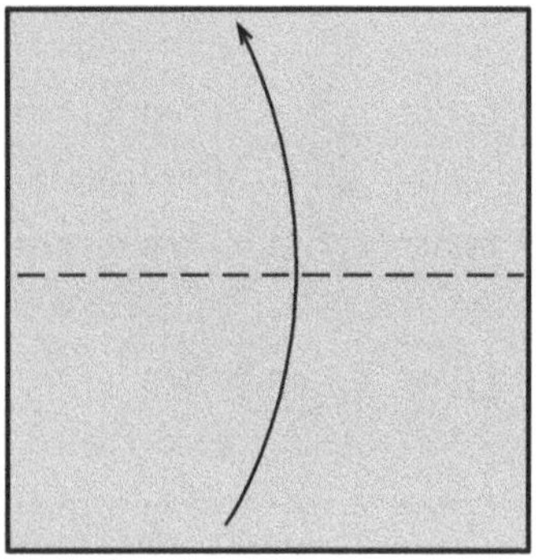

1 Fold in half.

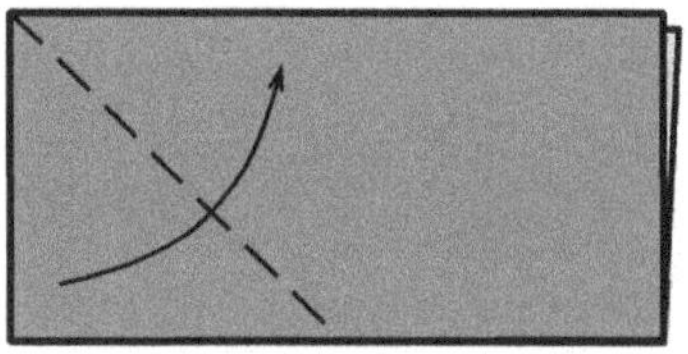

2 Fold the corner.

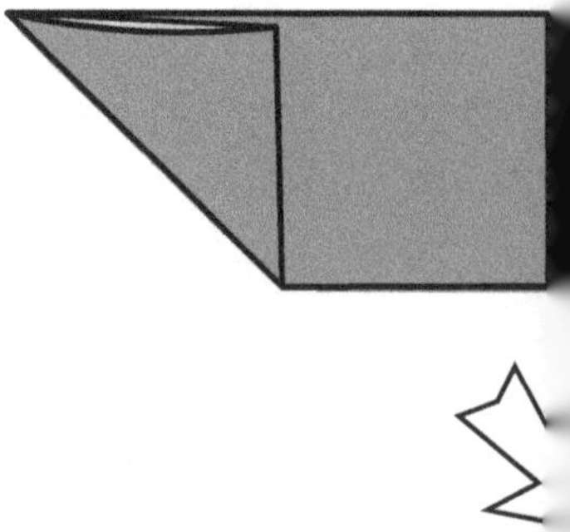

3 Unfold.

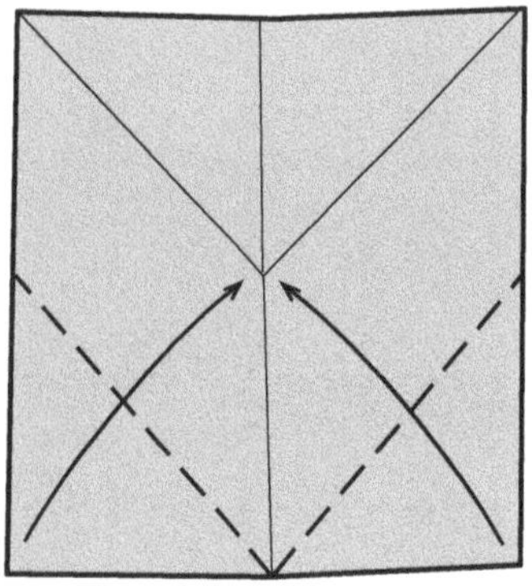

4 Fold the corners without creases to the centre.

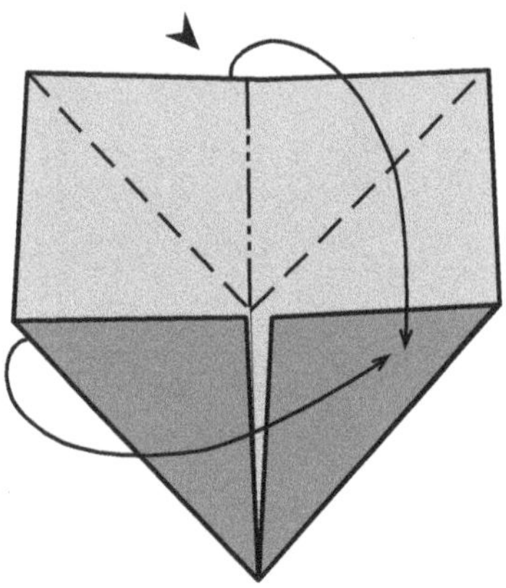

5 Make both partial diagonals valley folds. Use these folds to make a reverse fold as you close the paper in half.

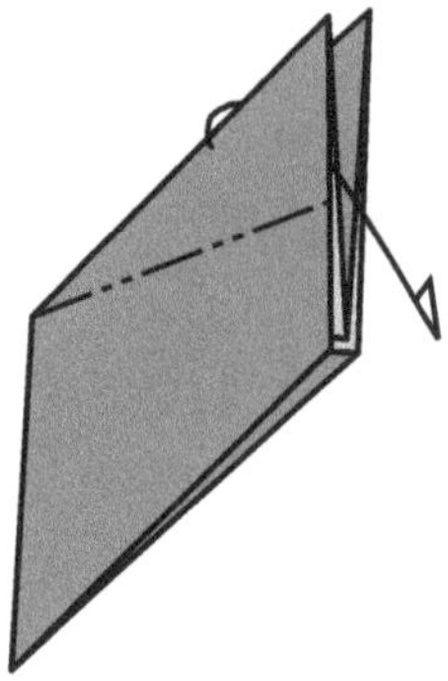

6 Fold the front flap insid
Leave the rear flap unfc
You need eight units; tv
four colours works well

ssembly ★

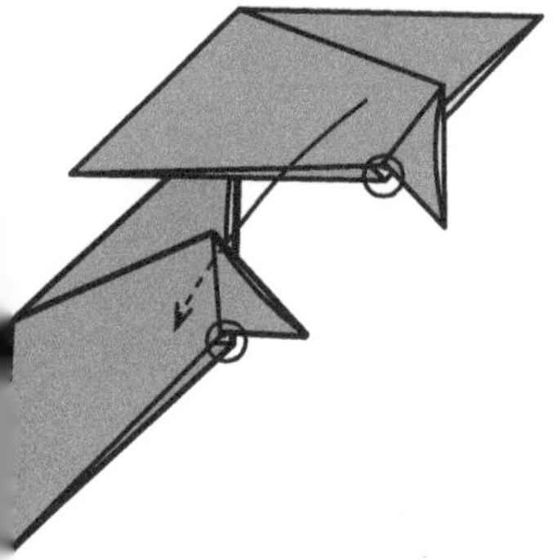

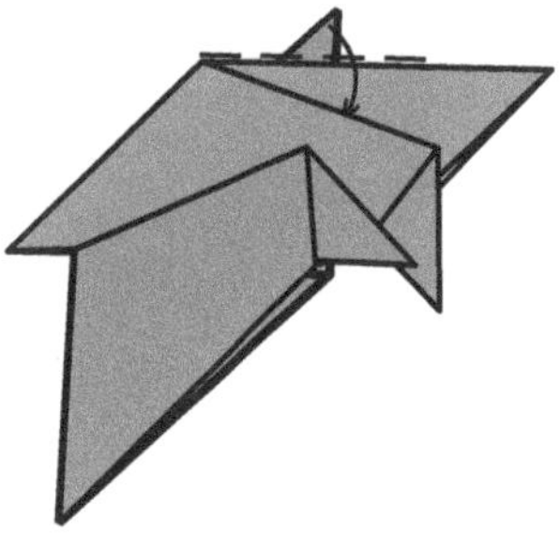

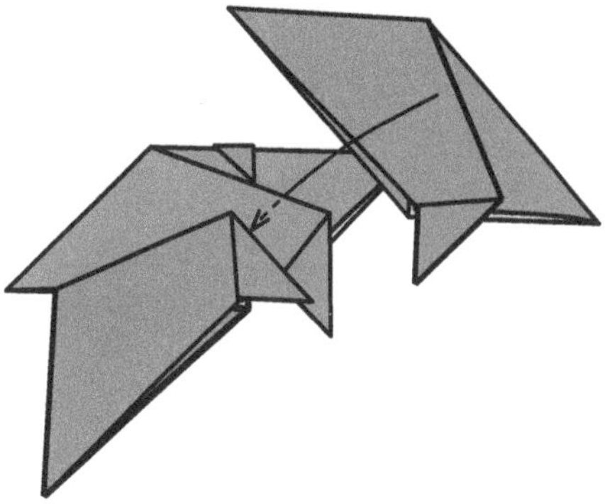

lip the second unit into he arms of the first unit. he circled points will be ne centre of the ring.

2 Fold the tip of the first unit over the second to create the sliding mechanism. Leave a small gap so that units slide easily.

3 Continue adding units in a clockwise direction: slip the third unit into the arms of the second unit.

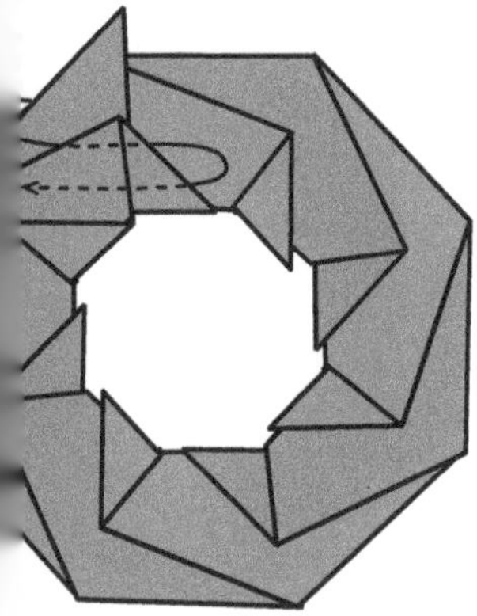

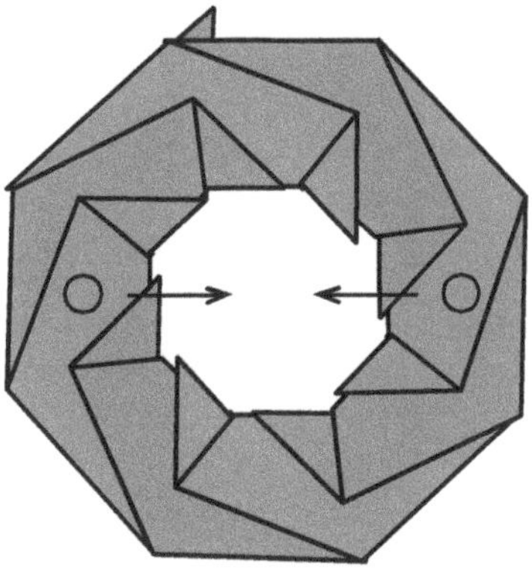

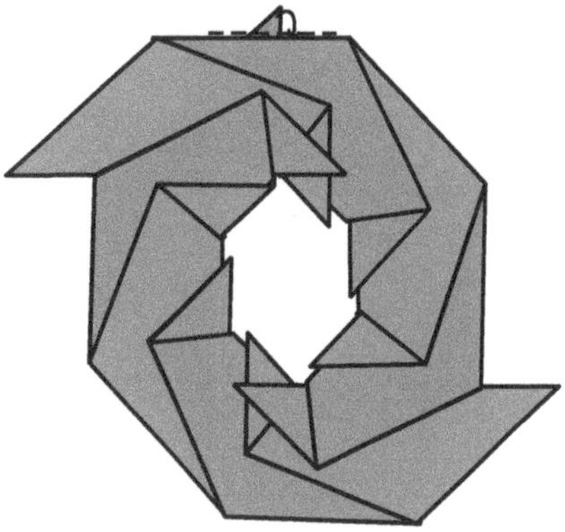

o make adding the last unit asier, remove the first unit nd add this as your new last nit. To make this helix into ring, flip the start of the ng towards you and tuck it to the arms of the last unit. arefully bend the ring to o this.

5 Whenever adding a unit or tucking a tip is difficult, slide the existing units to make the space you need.

6 Tuck the tip of the last unit into the first.

ction

ide the units by pushing them towards, or pulling away from, the centre. The ng will not close if you push the units too far: slide the ring open and try again.

Harlequin Slider

★★★☆☆

Use 4 square

Finished model ratio 1.(

This is another slider inspired by Robert Neale's classic Pinwheel-Ring-Pinwheel. 75 mm du origami paper works well: it has different colours on each side.

Module ★★★

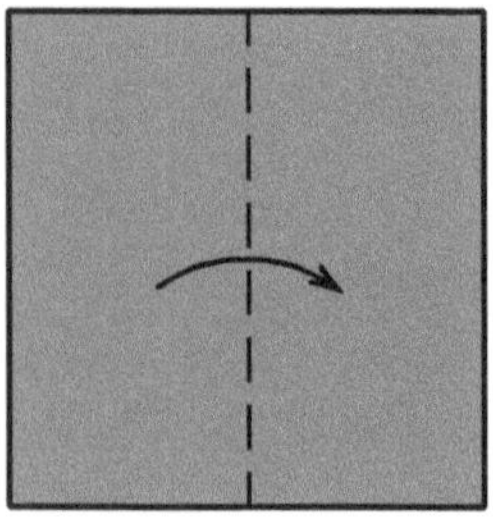

1 Fold in half.

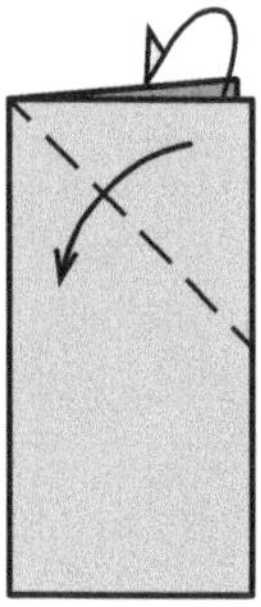

2 Fold the top corners down on the front and back sides.

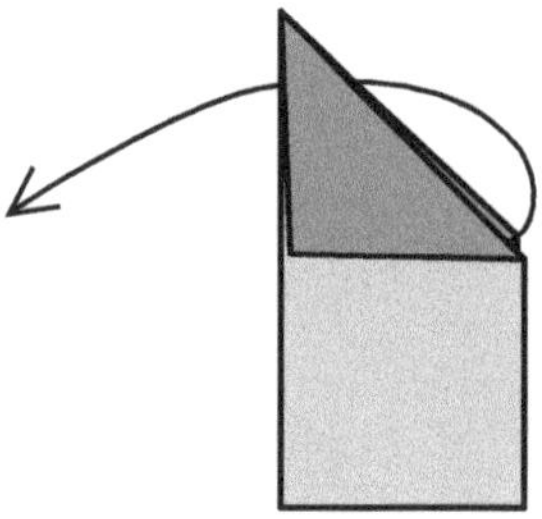

3 Unfold from behind.

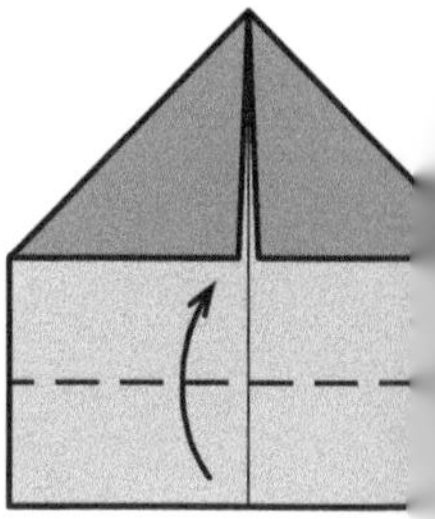

4 Fold the bottom edge up one quarter.

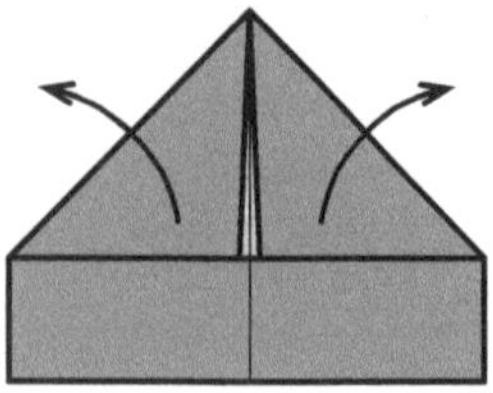

5 Unfold the corners.

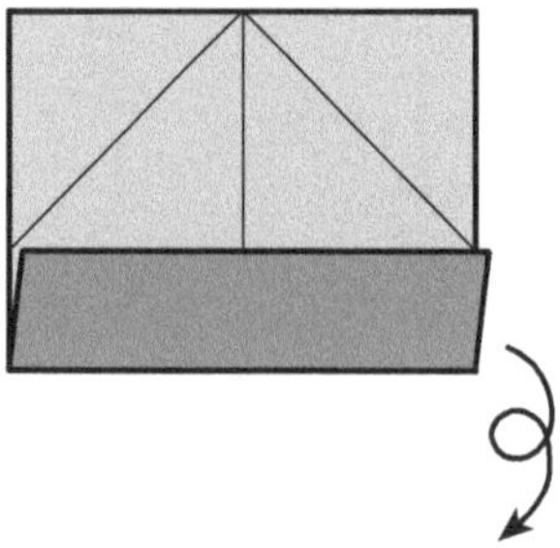

6 Turn over.

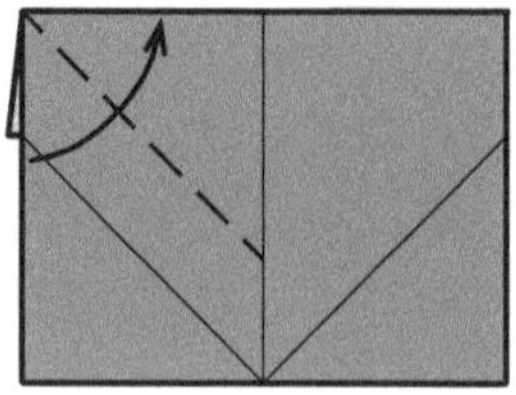

7 Fold the left corner up to the top edge. Only crease to the middle.

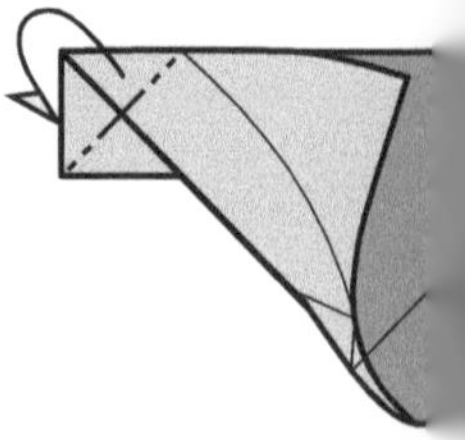

8 Fold the left cor behind.

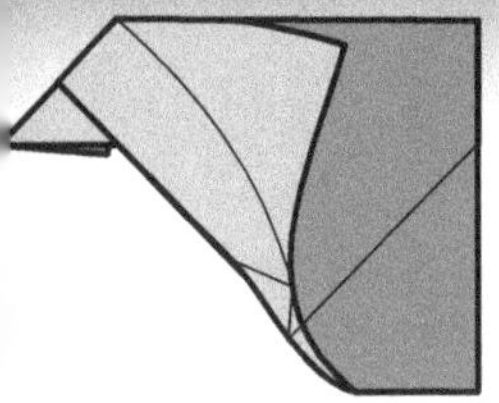

Unfold and repeat on the right. Then unfold completely.

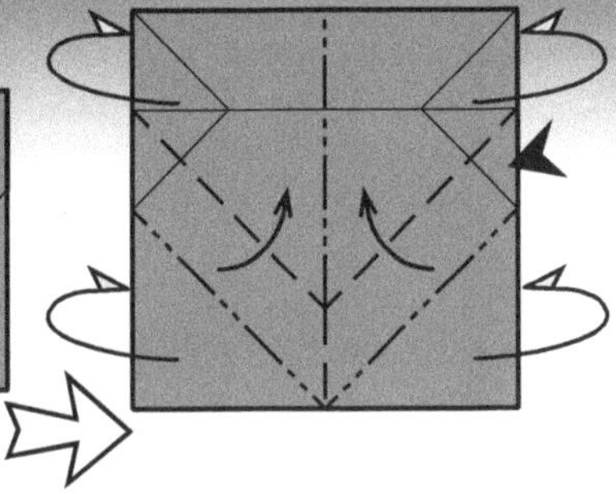

10 Make a pair of reverse folds, closing the model in half.

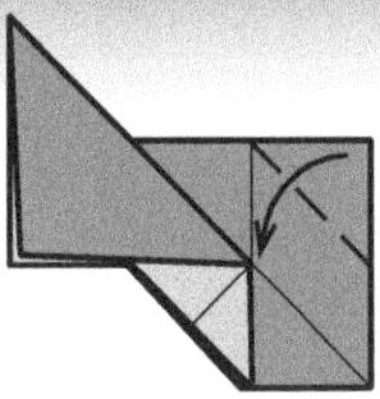

11 Fold the right corner down.

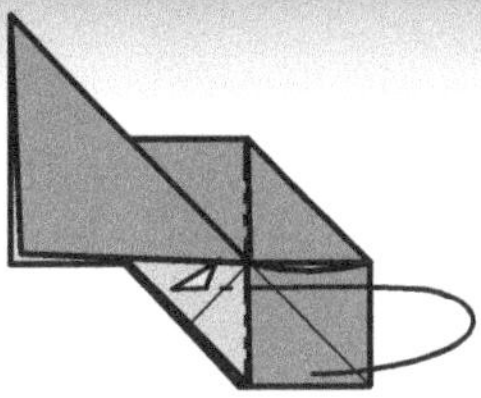

12 Mountain fold inside.

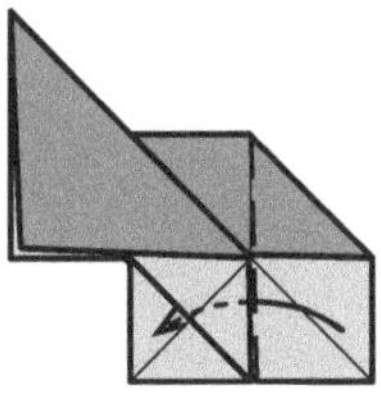

Fold inside.

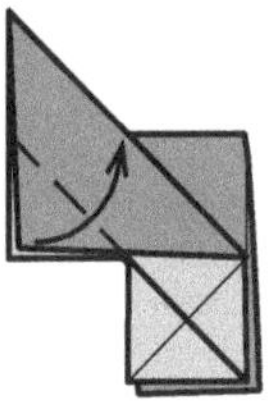

14 Fold the corners on the front and back.

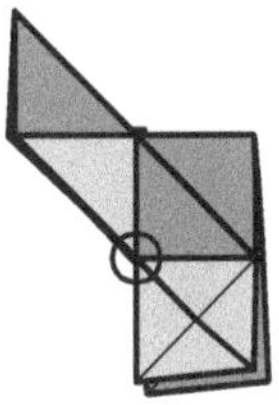

15 Make a total or four units. The circled point will be the centre of the finished model when it is in its smallest configuration..

ssembly ★★

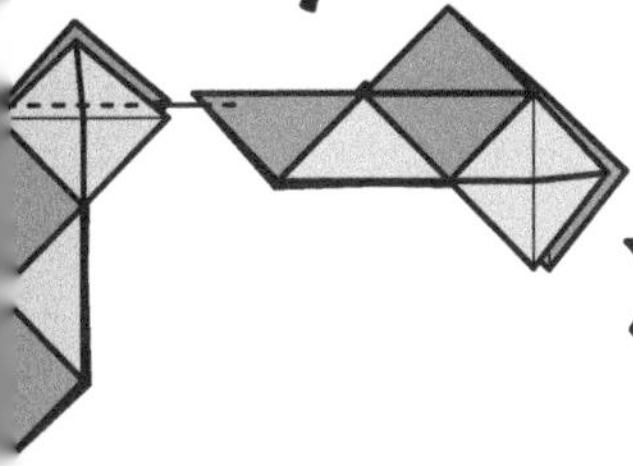

Slide the second unit into the arms of the first.

2 Mountain fold the tips of the arms into the second unit.

3 Add more units.

4 You may need to slide when adding the last unit.

Action

Hold opposite modules and slide. Do not slide fully open or it will be hard to close. f you want to avoid this, or are using paper that is white on one side and coloured on the other, make mountain folds in step 14.

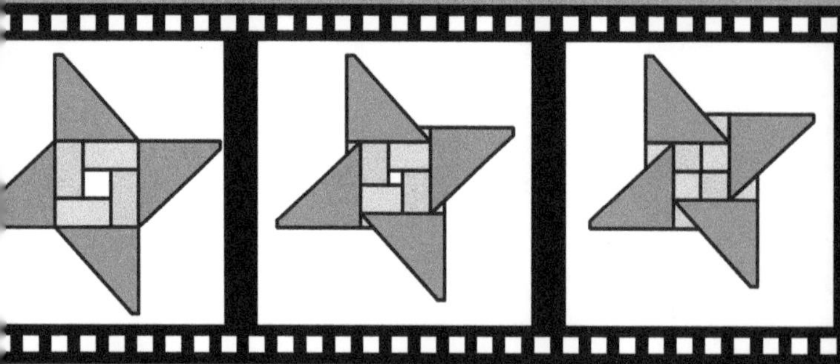

Shaky Shuriken Star

★★★☆☆

Use 4 squares

Finished model ratio 1.0

This is another slider inspired by Robert Neale's classic Pinwheel-Ring-Pinwheel. Here the "arms" are put in a pocket but are not fully trapped. The mechanism is like the "sliding dovetail" carpentry joining technique.

Module ★★★

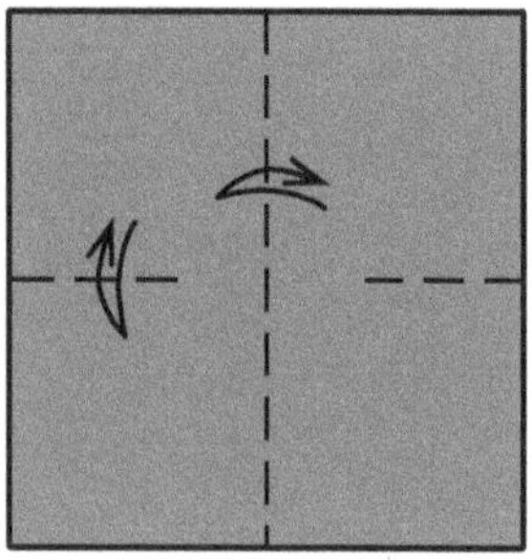

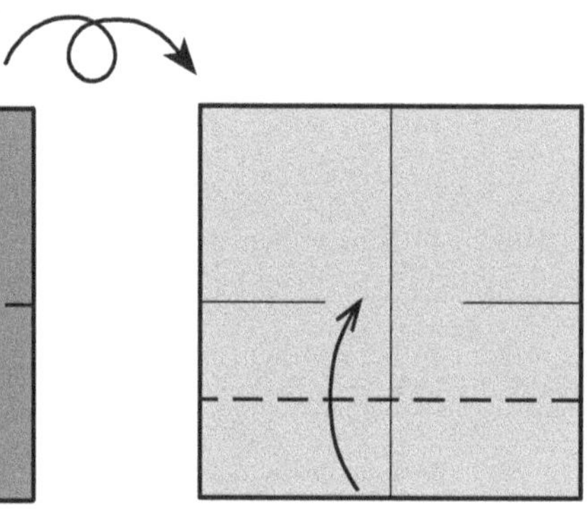

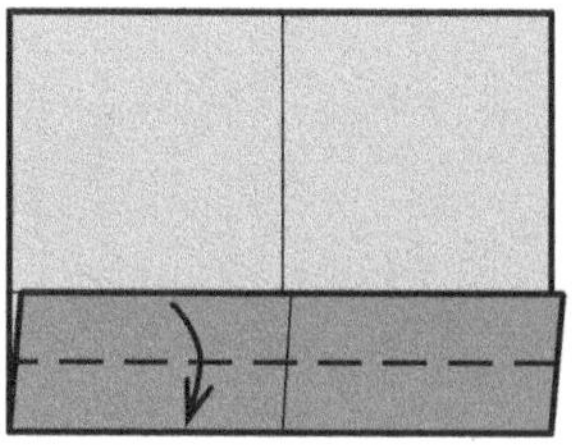

1 Make two folds in half. Do not crease the middle for the horizontal fold.

2 Fold the bottom edge to the middle.

3 Fold the top flap in half downwards.

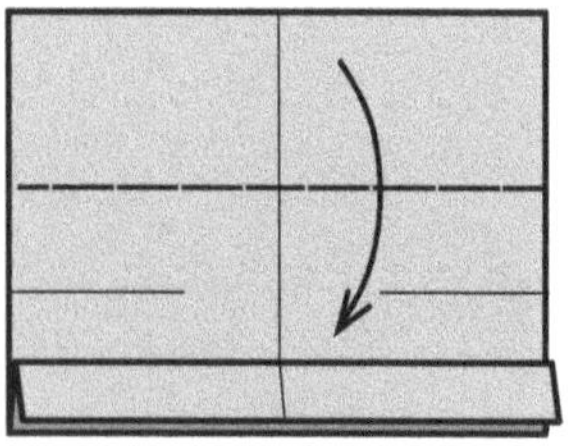

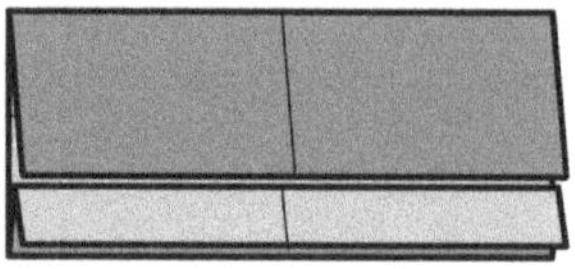

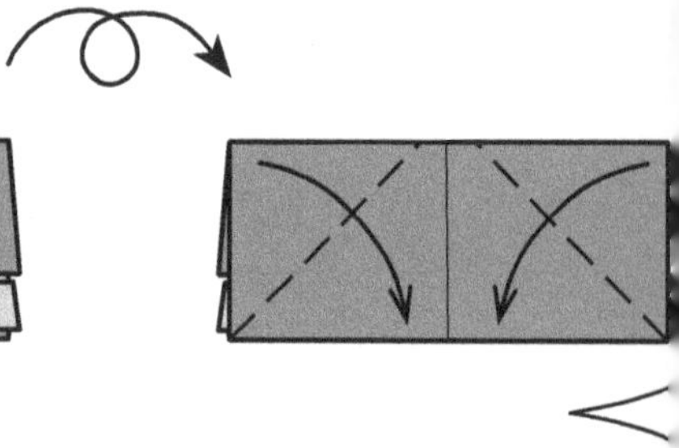

4 Fold the top edge to the folded edge that you have just made.

5 Like this. Now turn over.

6 Fold the left and right corners down, folding through all layers. There will be a gap in the centr

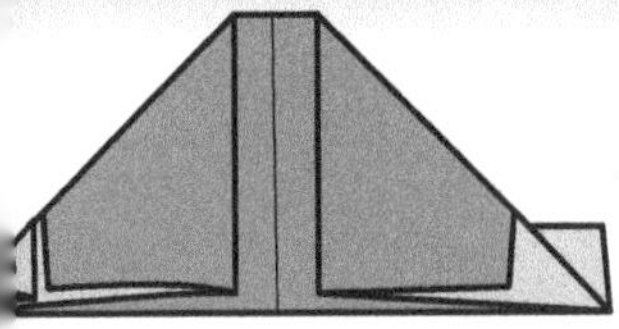

Unfold the small triangular flap on the left side.

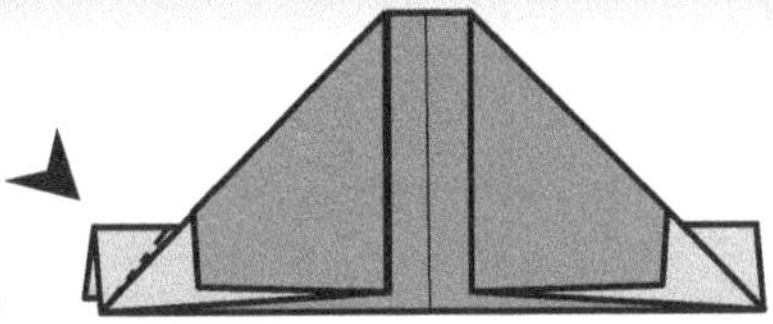

8 Reverse fold using the existing crease.

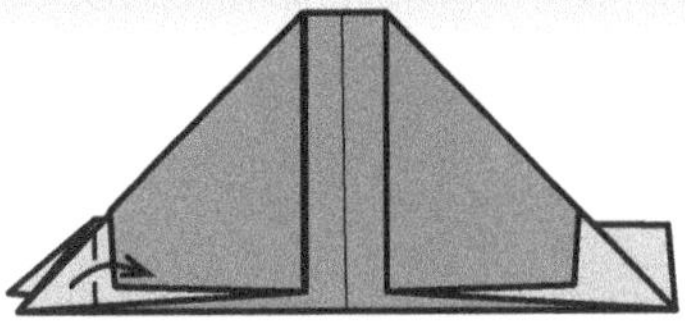

9 Fold to the right the little flap that you have just made.

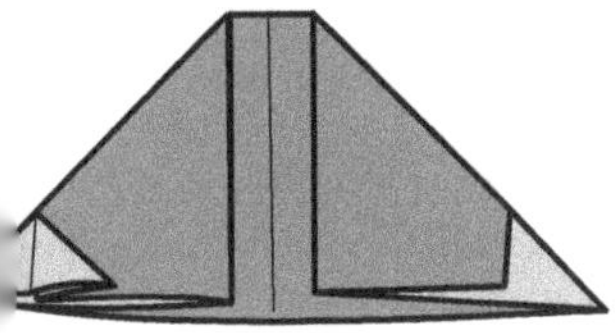

Like this. Repeat steps 7 to 10 on the right side.

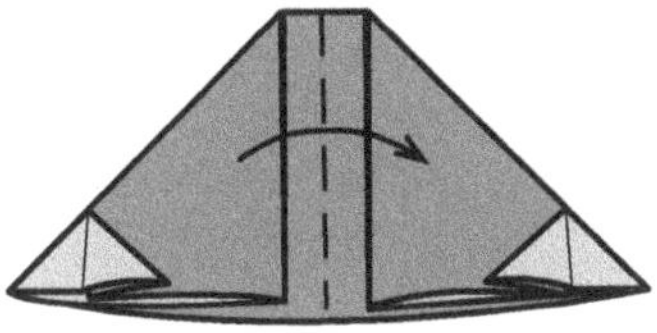

11 Fold in half using the vertical crease in step 1.

12 Make four units in total.

ssembly ★★★

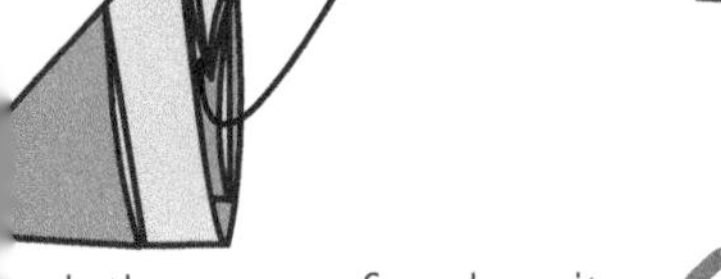

uck the arms of each unit to the pockets of the next nit. You can gently bend the ceiving unit to help tuck the ms in.

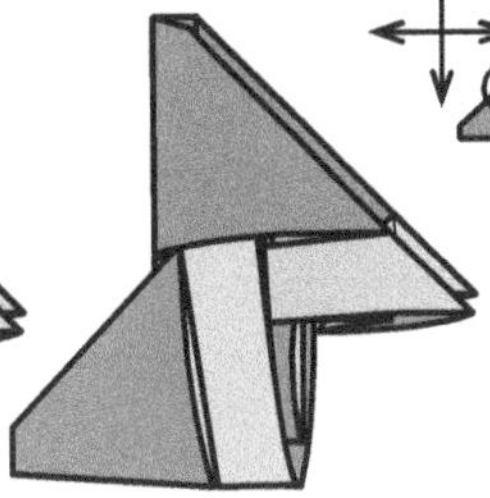

2 Repeat on the other side and continue adding units.

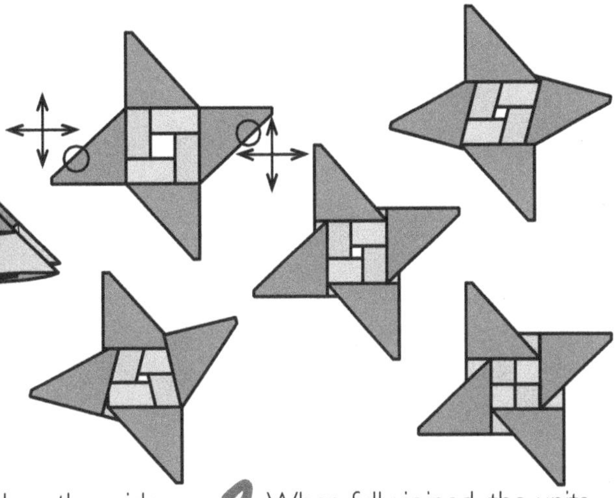

3 When fully joined, the units can take many positions. Here are some examples.

ction

old the circled points and push and pull. The maximum and minimum positions are own. What other shapes can you make?

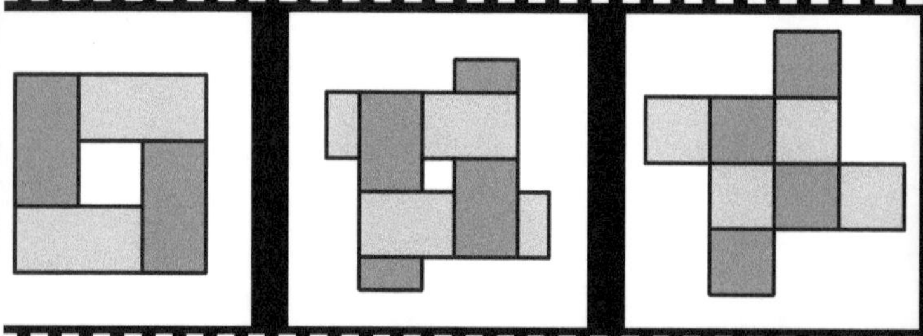

Square Slider

★★★☆☆

Use 4 square:
Finished model ratio 1.:

This slider is inspired by Robert Neale's classic Pinwheel-Ring-Pinwheel. Francis Ow and I exchanged different types of sliders. His used a waxy paper with each unit's "arms" sliding inside the next unit; these are less likely to come apart. I chose to have the arms on the outsi like Robert Neale's as they are easier to adjust, even if they can come apart more easily. Altho this model seems to be mundane as the inner hole can only be rectangles, you can continually slide in cycles without letting go and gripping the other units.

Module ★★★

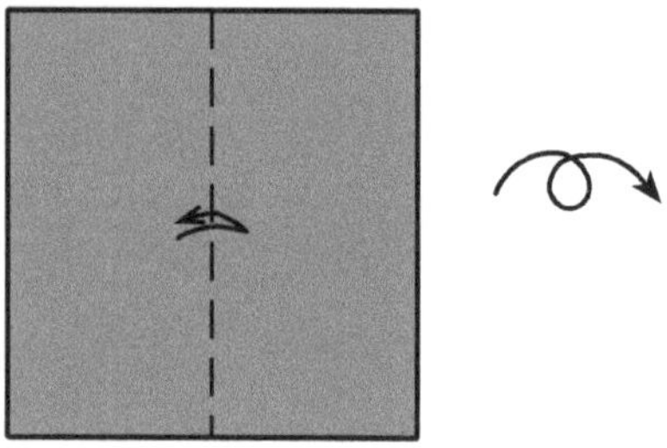

1 Use four 75mm squares. Crease in half and turn over.

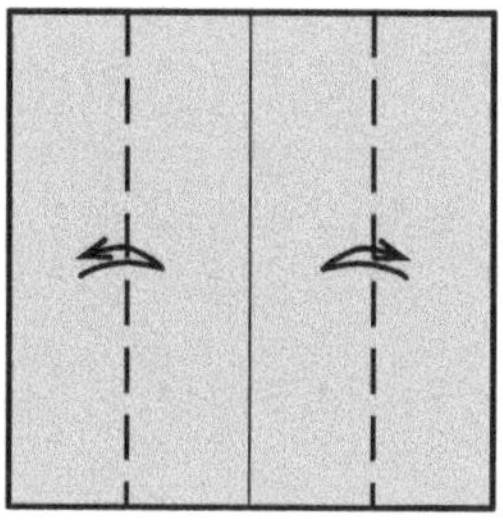

2 Fold the sides to the centre and unfold.

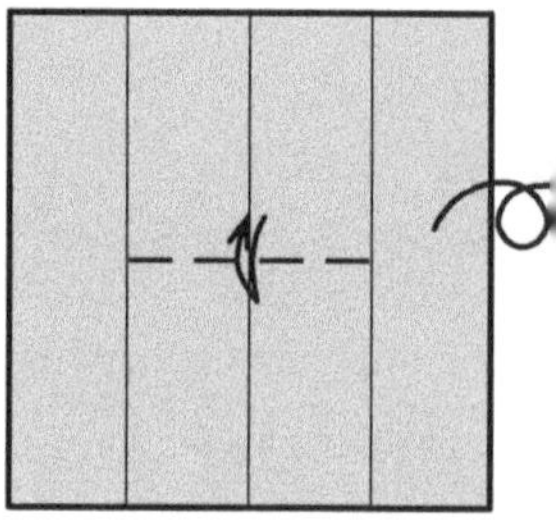

3 Fold in half only betwee the crease lines. Turn o

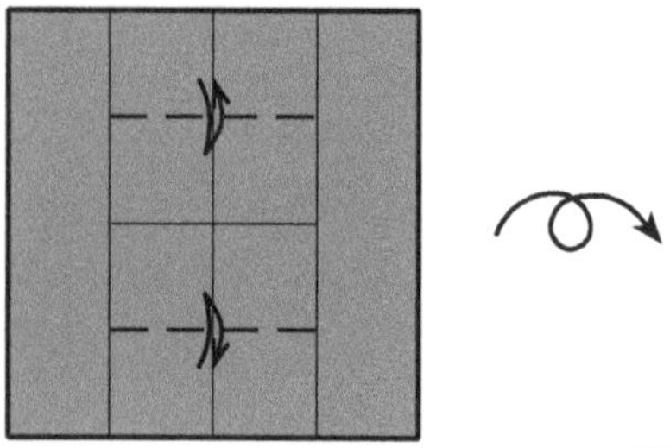

4 Fold quarters only between the crease lines. Turn over.

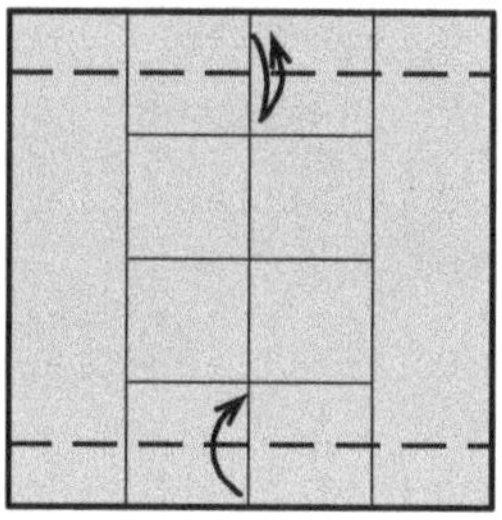

5 Fold the top and bottom edges inwards one eighth. Unfold the top edge.

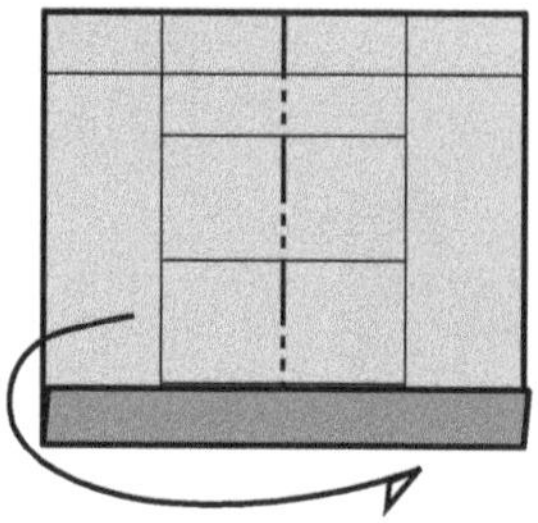

6 Fold in half behind.

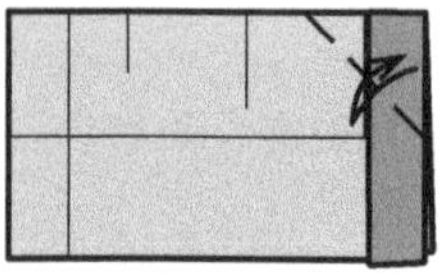

7 Crease the top right corner.

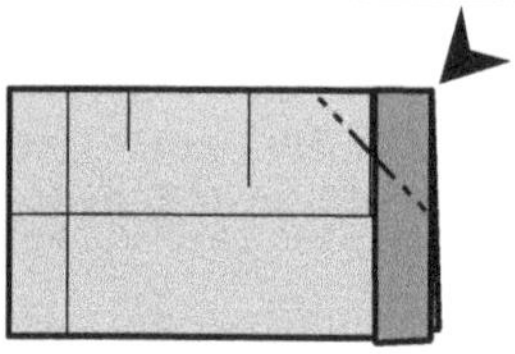

8 Inside reverse fold.

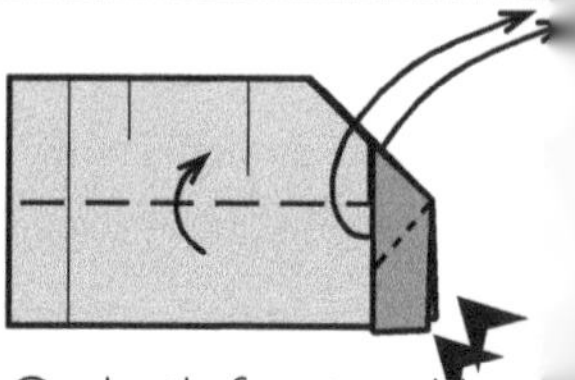

9 On both front and bac open the pockets and l the bottom edges upw

10 Fold the left edge inside. This will block the slider to stop it sliding apart.

11 Make four units in total.

Assembly ★★

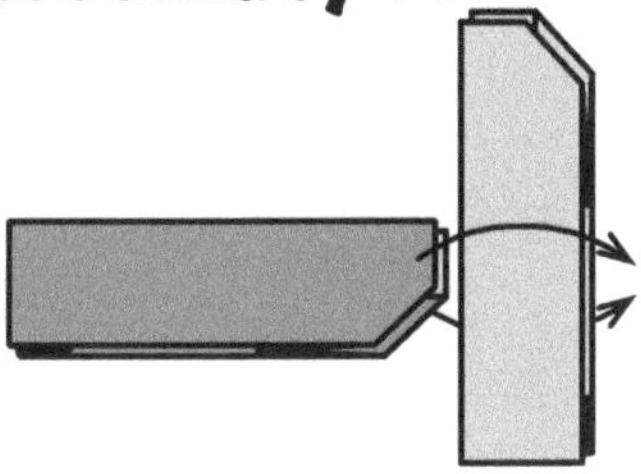

Wrap the arms of the first unit around a second unit. To orientate the units, note that the centres of the original squares meet at the centre of the slider.

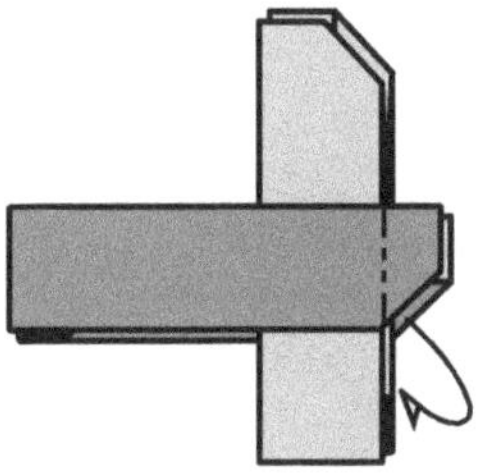

2 Tuck flaps in to lock. Test the sliding action and either tighten or loosen the arms, if needed.

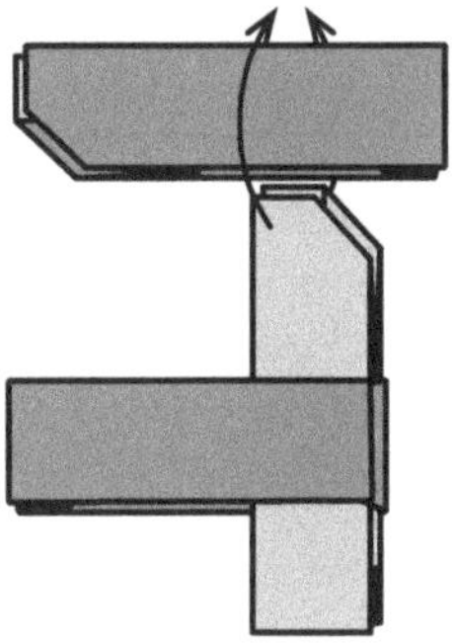

3 Continue adding units.

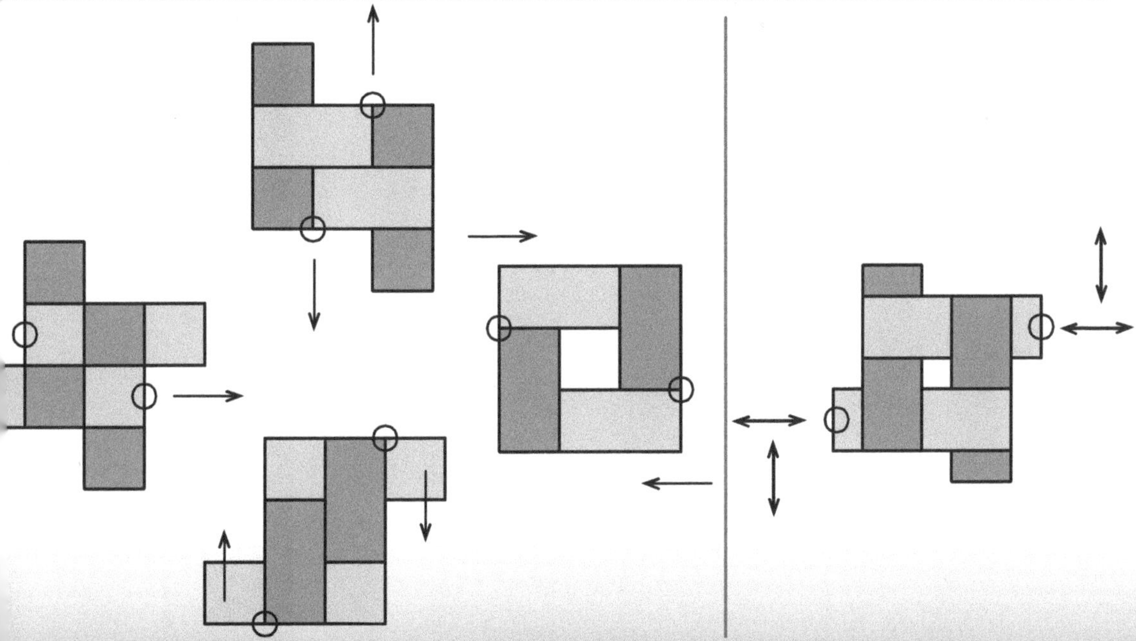

Action

Hold the circled locations and pull or push. To continually slide between intermediate positions, hold the circled locations and push or pull horizontally or vertically. Only apply forces in the plane of the slider or it will buckle.

Hex Slider

★★☆☆☆
Use 6 squares
Finished model ratio 1.7

This slider uses 60 degree geometry. This and the next slider have holes in their smallest position. This slider shows that the "arms" of unit can be a single layer of paper. This versio is not blocked — slide too much and the units come apart! 10cm squares of photocopy paper work well. You can use two or three colours.

Can you vary the angle and paper to make four- and eight-piece versions? How could you mak a version that closes without a hole?

Module ★★

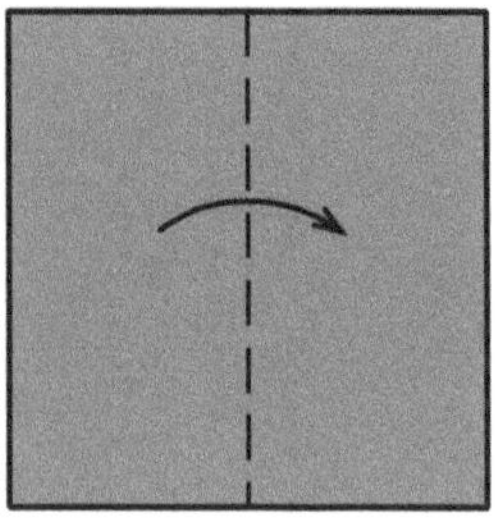

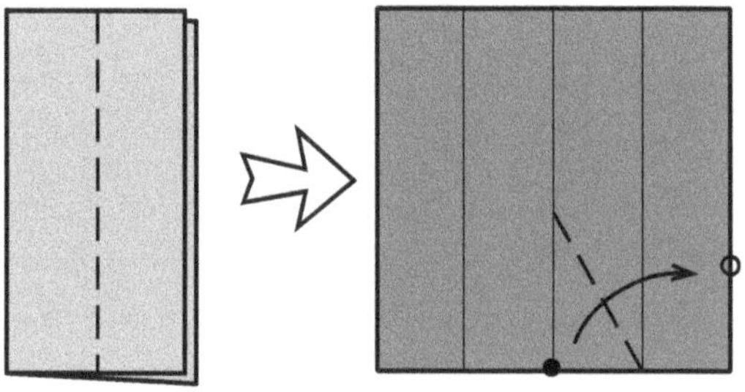

1 Fold in half.

2 Fold the raw edges to the folded edge. Unfold.

3 Starting from the bottom three quarter point, fold midpoint of the bottom edge onto the right raw edge. Only fold up to the middle crease.

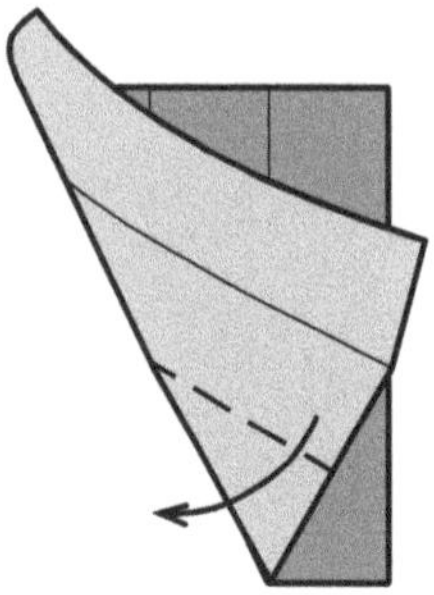

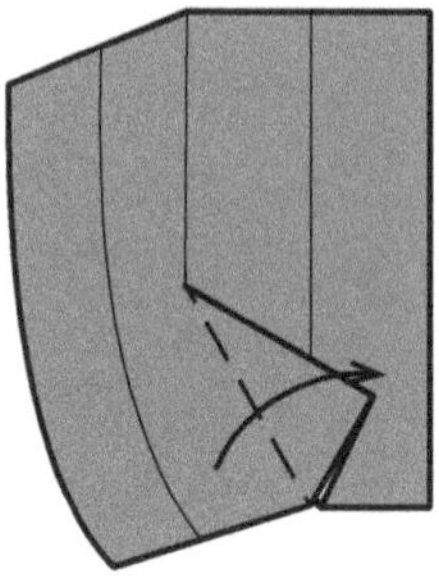

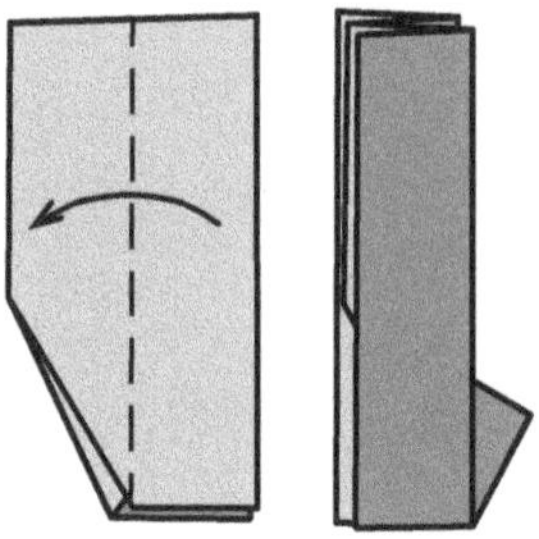

4 Fold back to create a concave shape.

5 Fold flat.

6 Fold on the quarter lines made in step 2.

ssembly ★★

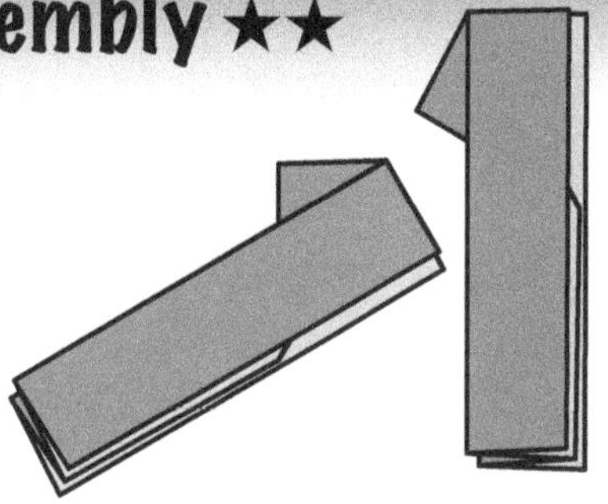

1 Put the second unit into the arms of the first unit.

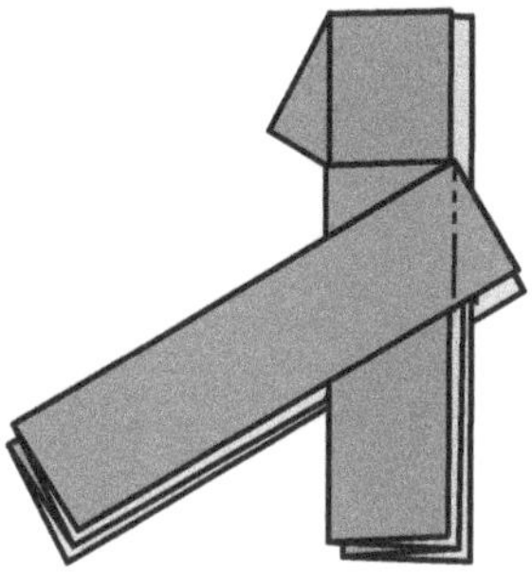

2 Fold the tips of the first unit's arms into the second unit. The tips lie between the raw edges and the folded edges.

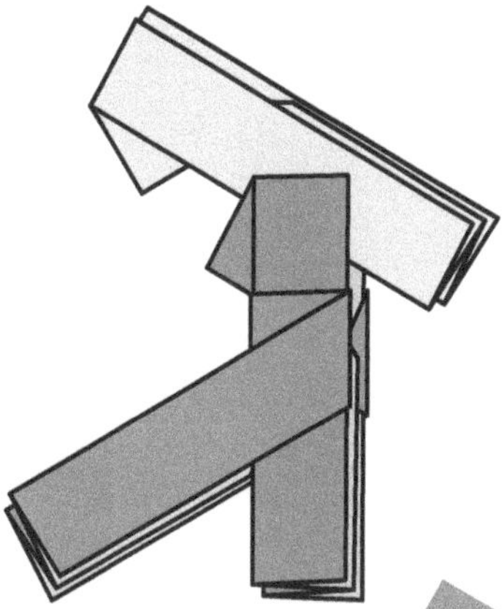

3 Add more units.

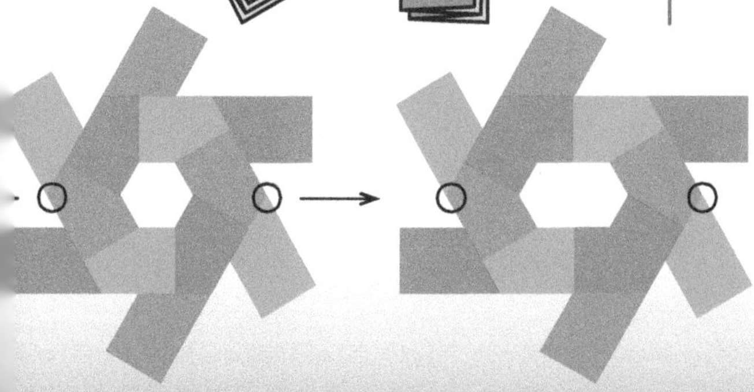

ction

old the circled locations and slide. To continually slide between intermediate ositions, hold the circled locations and push or pull. Only apply forces in the plane of e slider or it will buckle.

Ribbon Slider

★★★☆☆

Use 8 square

Finished model ratio 1.

This model uses a sliding dovetail. The size of the ribbon determines the size of the dovetail. A big sliding dovetail is easier to fold and less likely to pop out when sliding. However, it is harder it is to assemble and does not slide as far as a smaller dovetail. The proportions have been carefully chosen to balance these features. This model will also spin on a flat surface when it in its smallest configuration.

Module ★★★

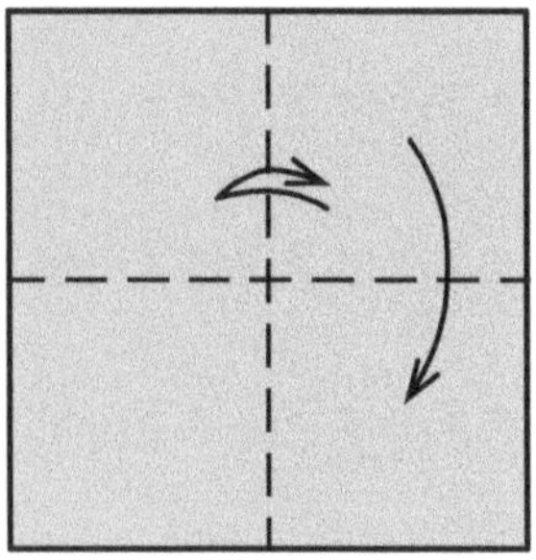

1 Crease in half and then fold the top down.

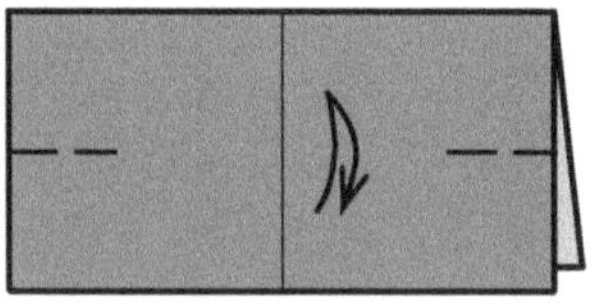

2 Without creasing the middle, fold the bottom edge to the top and unfold.

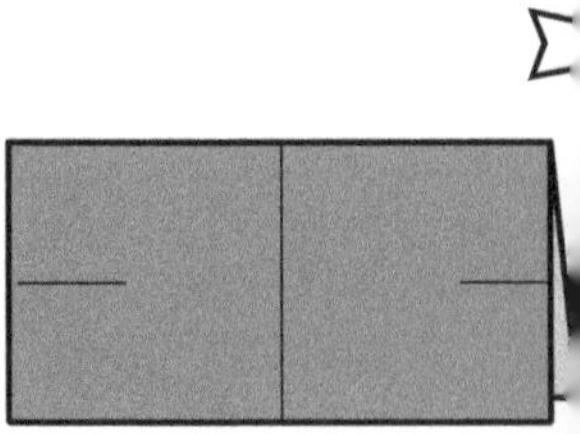

3 Unfold and return to st

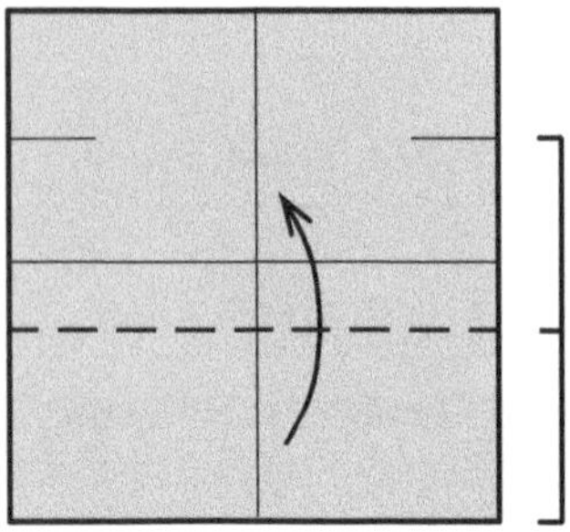

4 Fold the bottom edge up to the three quarter creases.

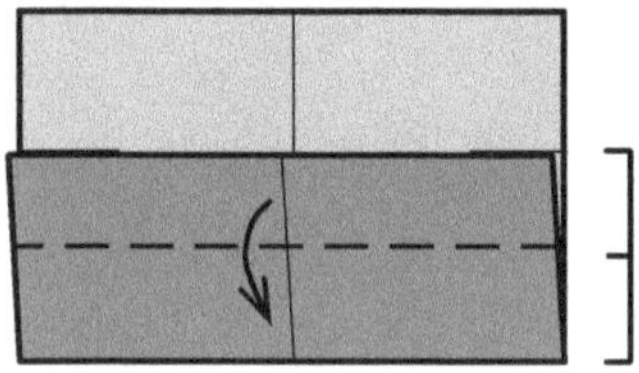

5 Fold the raw edge down to the bottom folded edge.

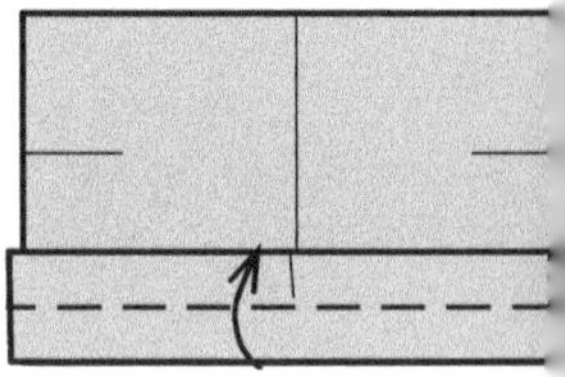

6 Fold the raw edge up t meet the folded edge.

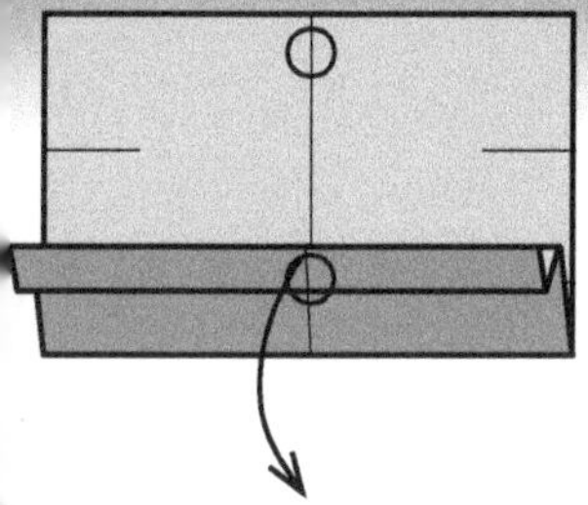

Hold the top two layers and pull down so that the top flap stays in place.

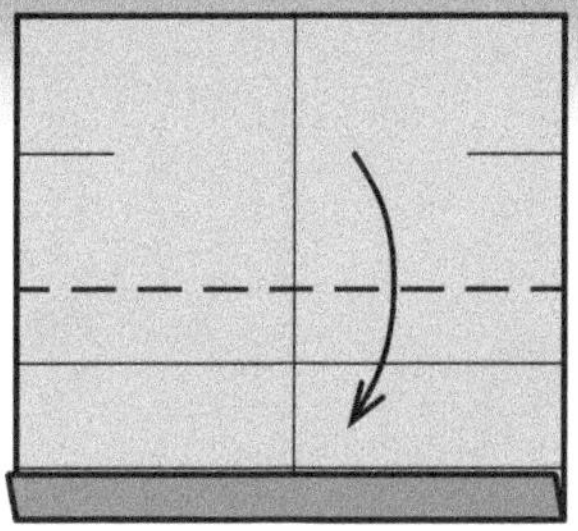

8 Fold down on the crease made in step 1.

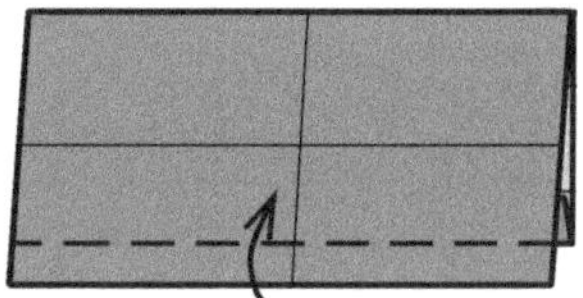

9 Fold the bottom edge to match the back flap.

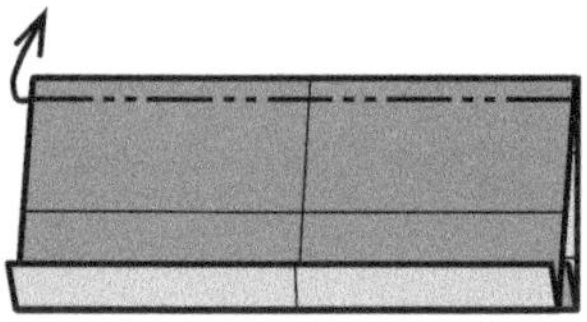

Slide the top two layers so that the bottom edge meets the raw edge behind. Make a new mountain fold.

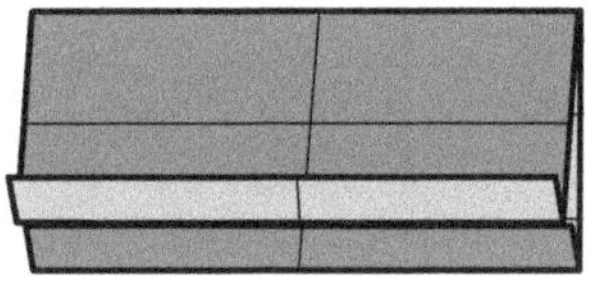

11 Turn over.

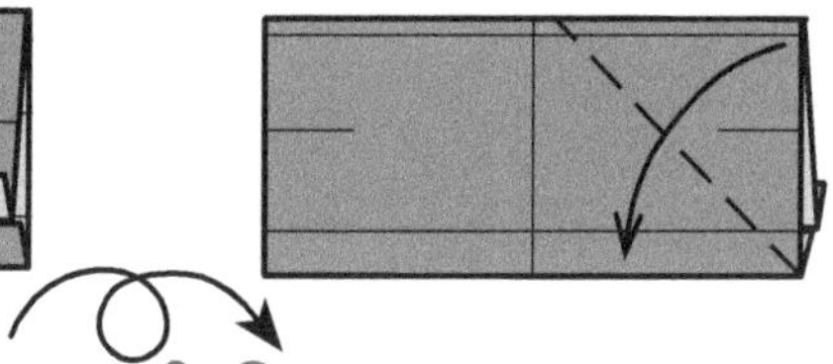

12 Fold the corner down, but do not crease the rear flaps.

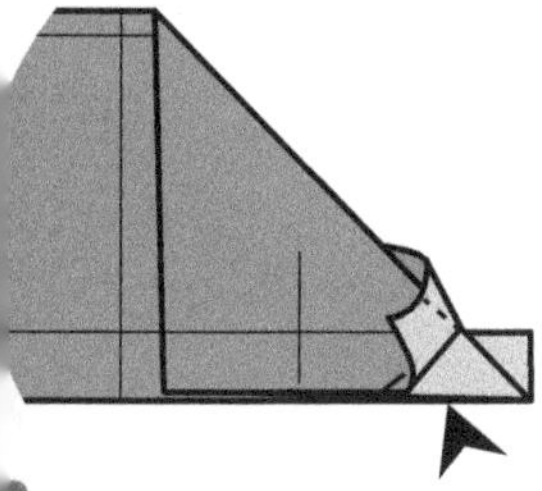

Squash the right flap.

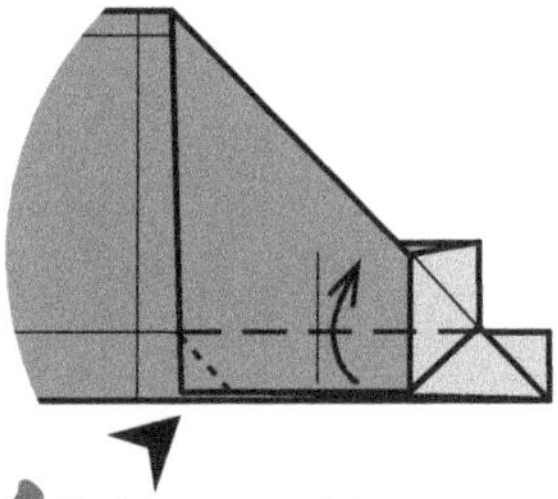

14 Fold up making a squash fold at the left corner.

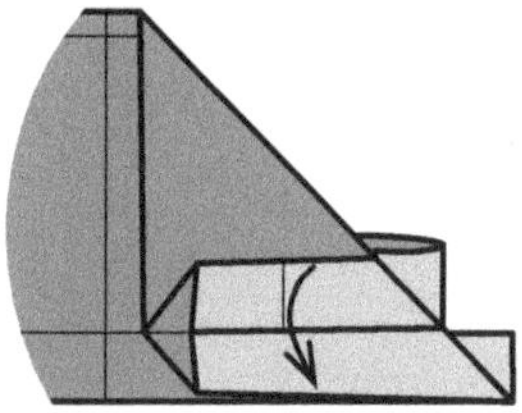

15 Fold down.

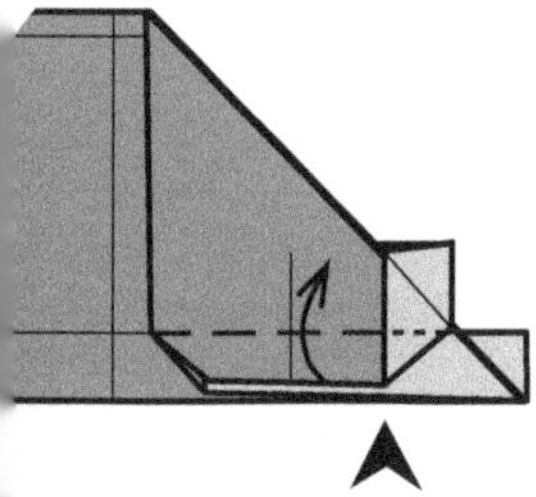

Fold up with an inside reverse fold at the right.

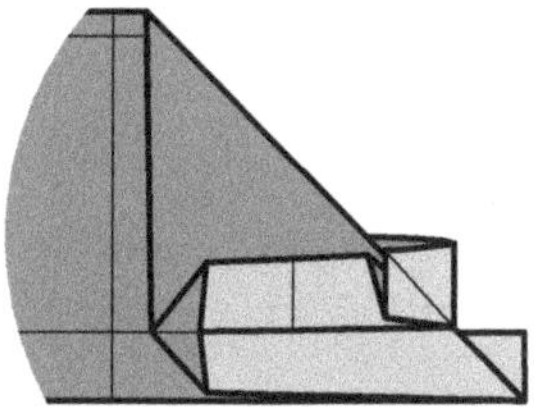

17 Repeat steps 12 to 16 on the left.

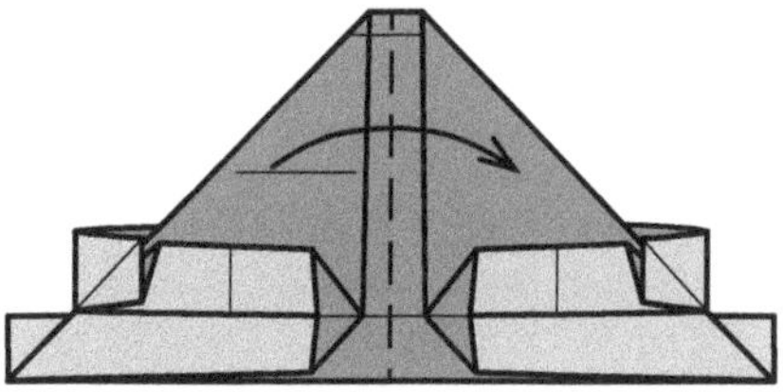

18 Fold in half to finish. Make eight units in total.

Assembly ★★★

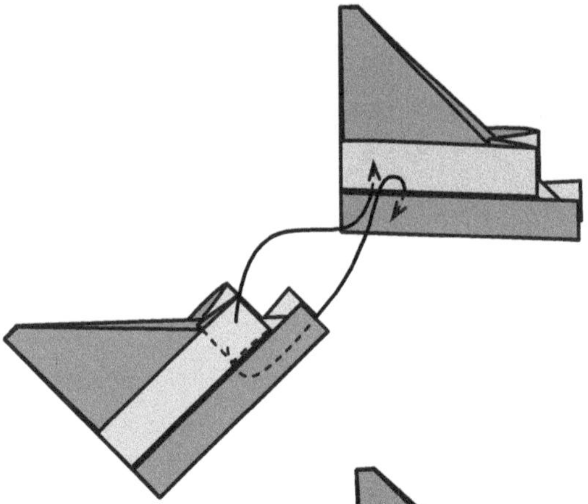

1 Tuck the sliding dovetail into the pocke[t] of the next unit. You can gently bend the receiving unit to help tuck in the tip of the dovetail.

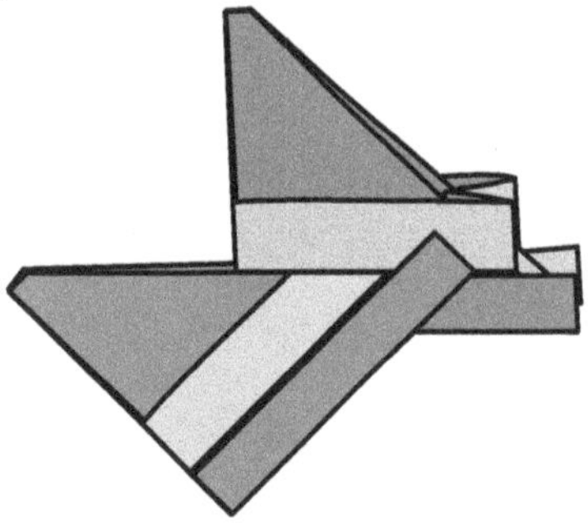

2 Repeat on the other side and continue adding units.

3 When fully joined, the units can take many positions. Here are some examples.

Action

Hold the circled points and push and pull. What shapes can you make? This model will also spin on a flat surface when it in its smallest configuration.

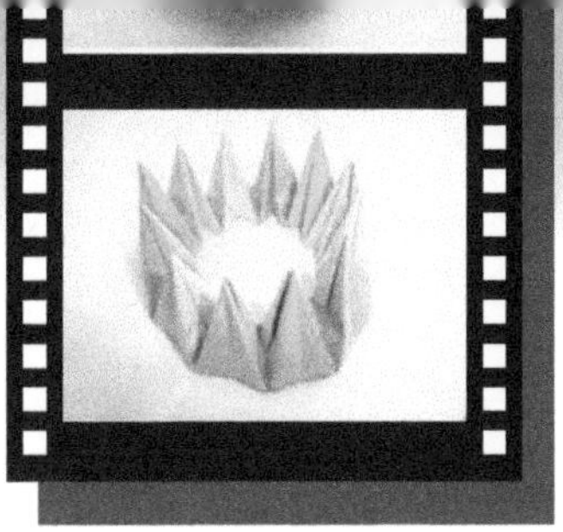

Chapter 2

Flexagons and Rotating Rings

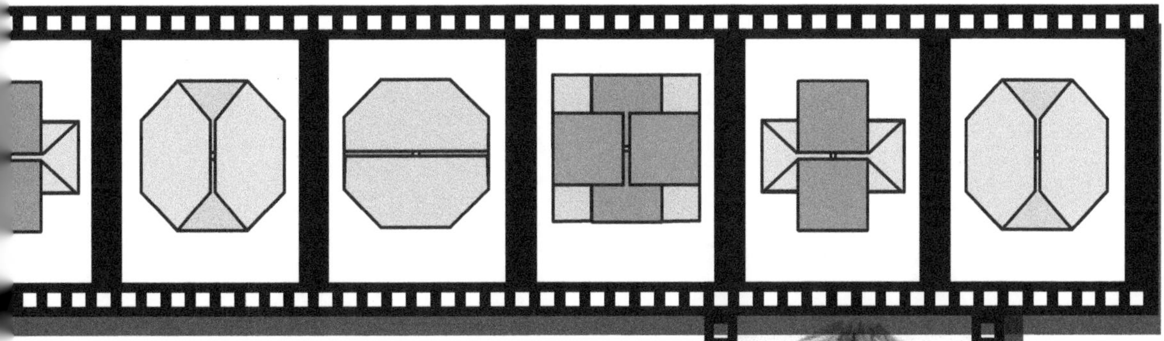

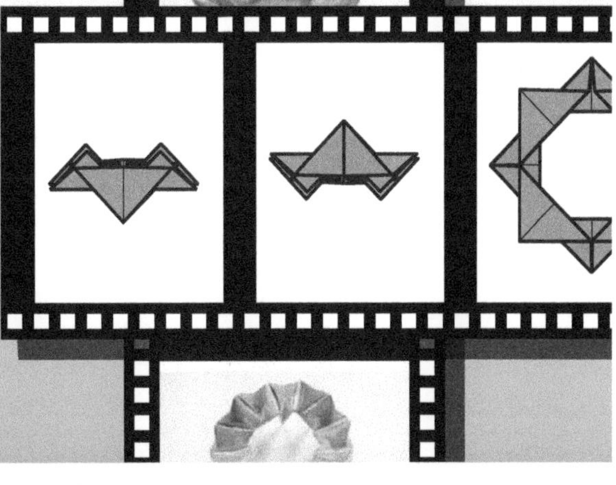

Three-piece Tri-hexaflexagon

★★★☆☆

Use 3 2:1 rectangle
Finished model ratio 1.2

Flexagons were first explored in depth by Arthur Stone and colleagues. Although beautifully simple, the original hexaflexagon requires taping or gluing. Challenged by Denver Lawson make a version without gluing, I found a couple of ways to lock the ends of the paper strip (o version is in *Learning Mathematics with Origami*). However, the locked triangle is much thicker the other triangles and can be distracting during flexing.

This three-piece modular version also has some thick triangles but has a satisfying "snap" whe flexing from concave to convex. The three colours move around the hexagon with sufficient flexing: this is not apparent in the one-piece version. The tri-hexaflexagon is a Möbius strip: a single side and a single edge.

Module ★★

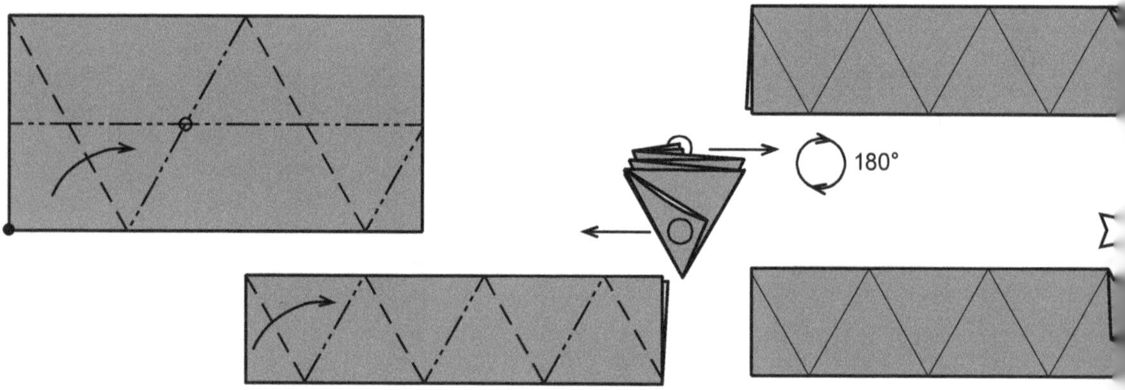

1 Fold the long edges together and unfold. Crease 60° folds. Mountain fold in half behind.

2 Starting at the left, crease 60° folds into a concertina.

3 Open and return to step 2. You may need to rotate the paper for the next step.

4 The folded edge at the top. Squa the top right co and then compl unfold.

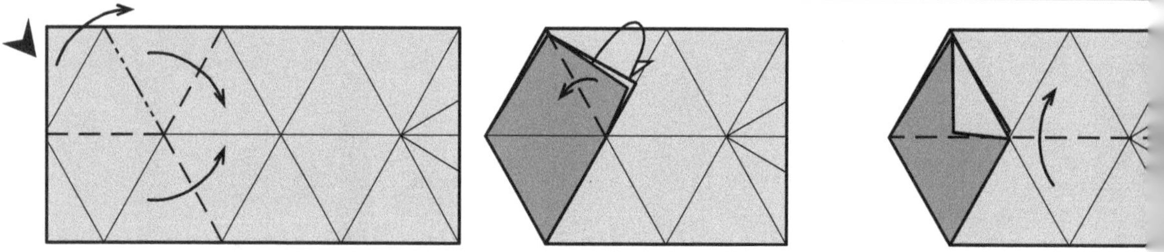

5 Using the existing creases, make a rabbit ear.

6 Fold the flaps. You do not need to tuck the mountain folded flap into the pocket.

7 Valley in half upwards.

sembly ★★★

Take two units. Return the second unit to the step 5, after the rabbit ear. Slide the tip of the second unit into the first unit.

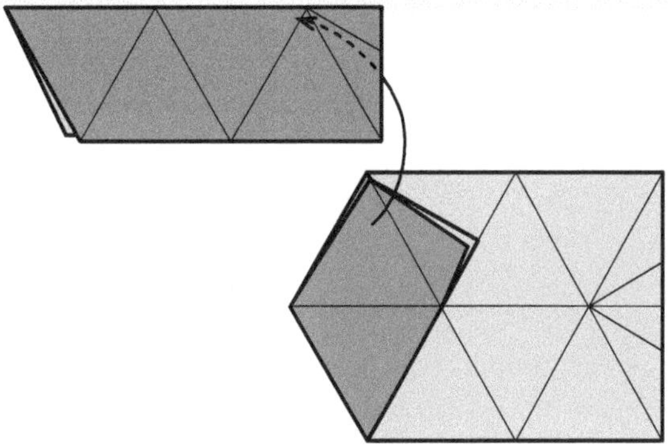

Reform the squash fold to trap the layers of the second unit

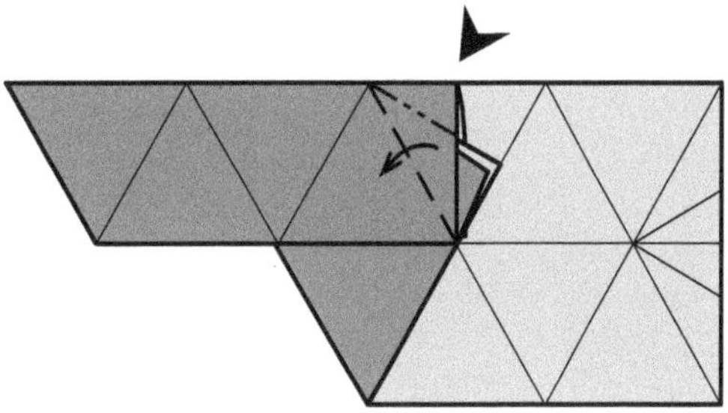

Tuck the right flap behind into the pocket of the second unit.

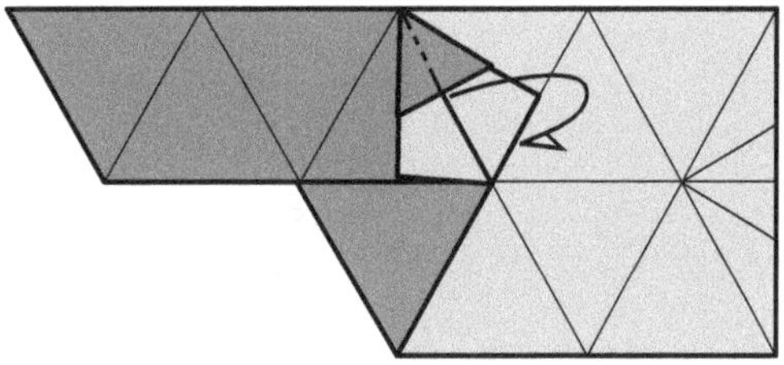

Fold the second unit in half. At the same time, tuck the flap of the first unit into the pocket of the first unit.

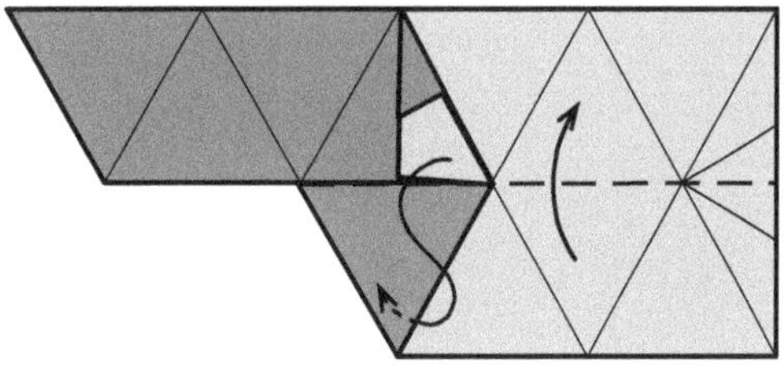

Turn over.

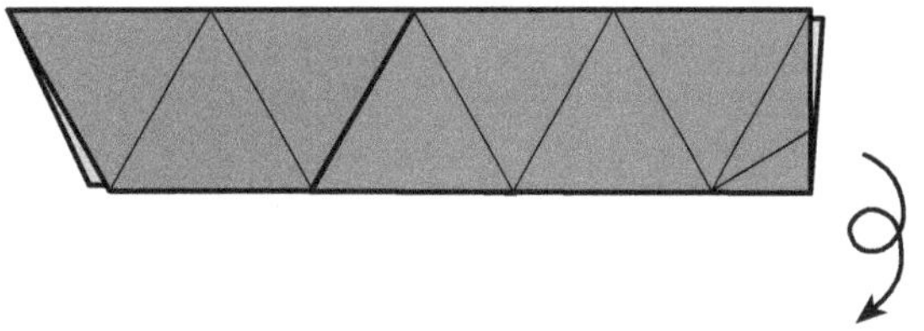

6 Add the third unit using the same joining method in steps 1 to 4.

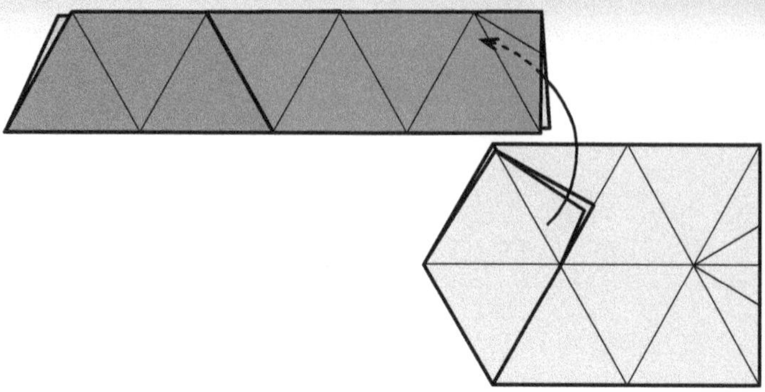

7 Turn over.

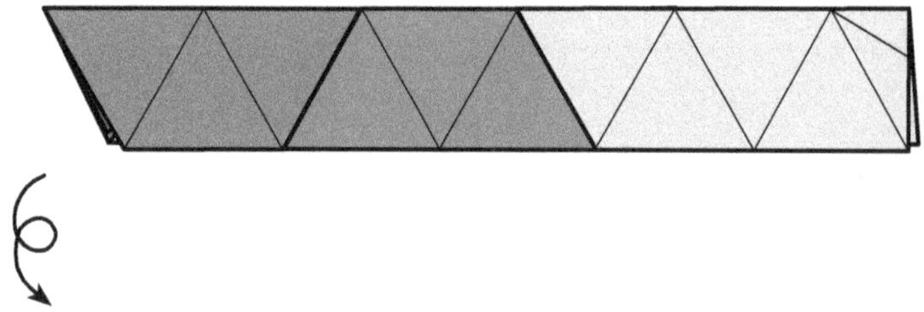

8 Pull out and unfold the flap at the left end. Mountain fold the right end.

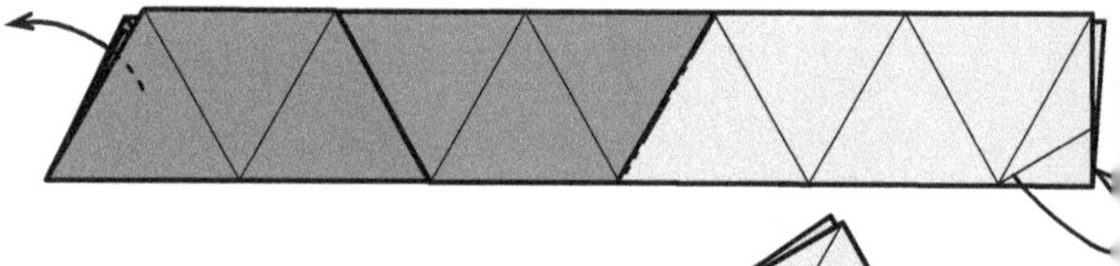

9 Valley fold the left end under the right end.

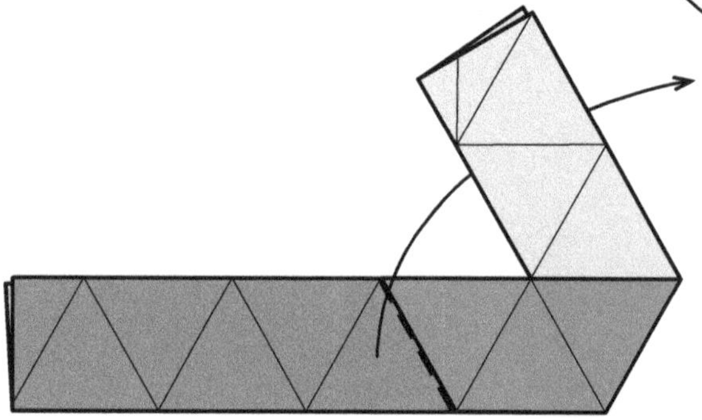

10 Tuck the flap inside.

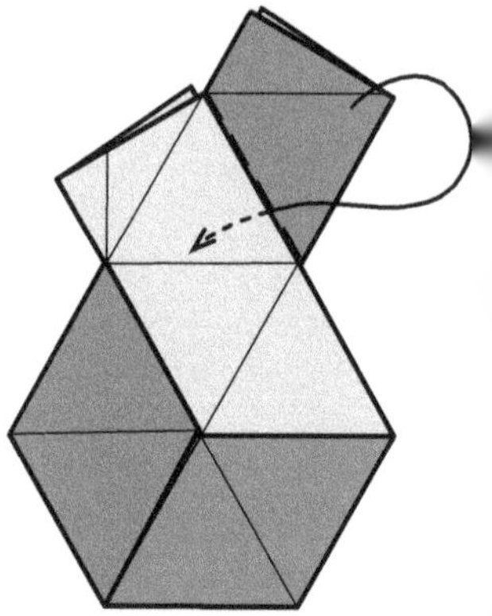

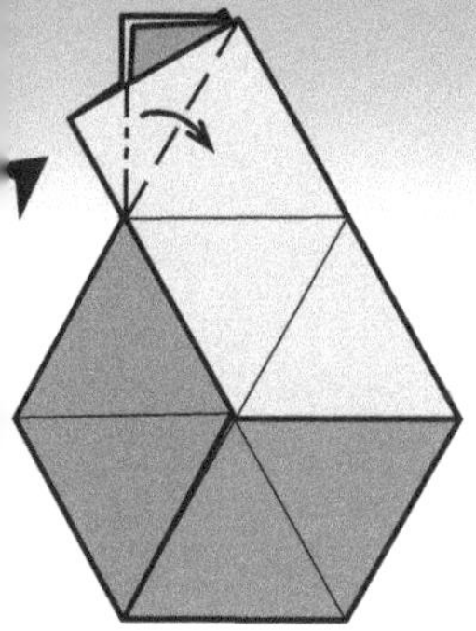

Squash the flap to trap the layers.

12 Fold the flap behind into the pocket.

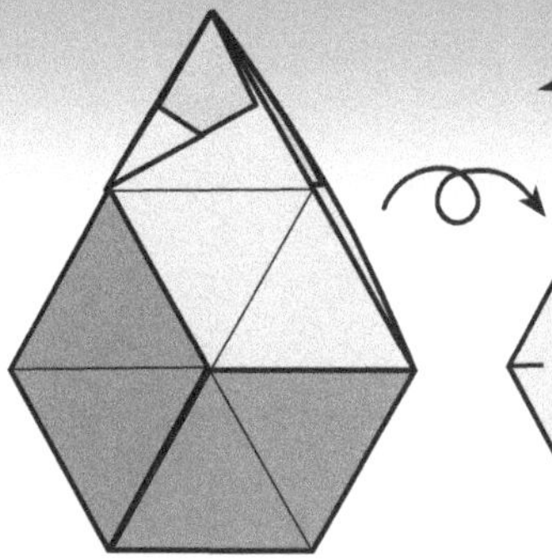

13 Turn over.

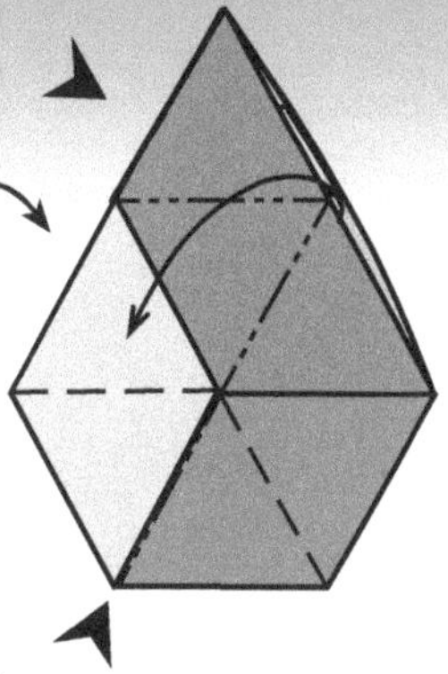

14 Open the top right pocket and squash. The shape is a partial octahedron.

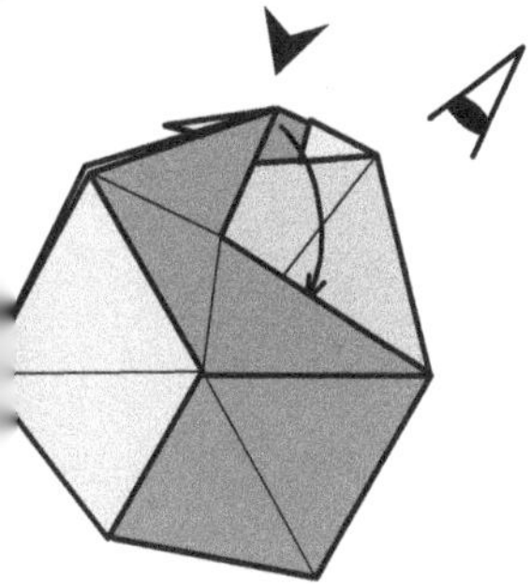

Push fully inside.

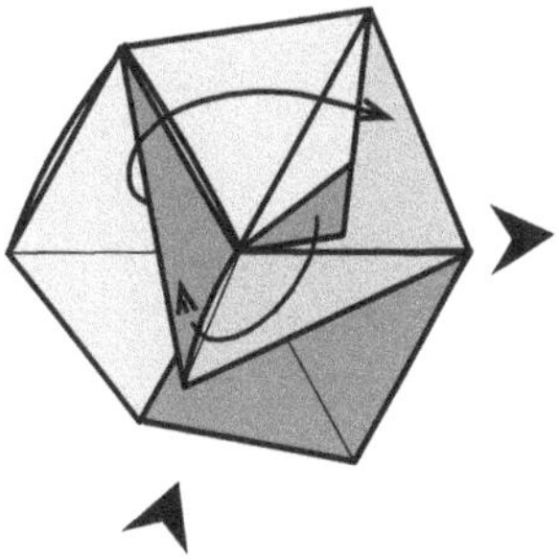

16 Tuck in while closing.

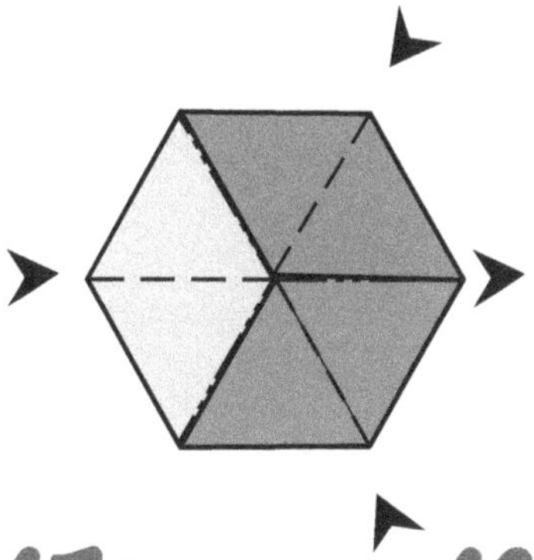

17 Squeeze.

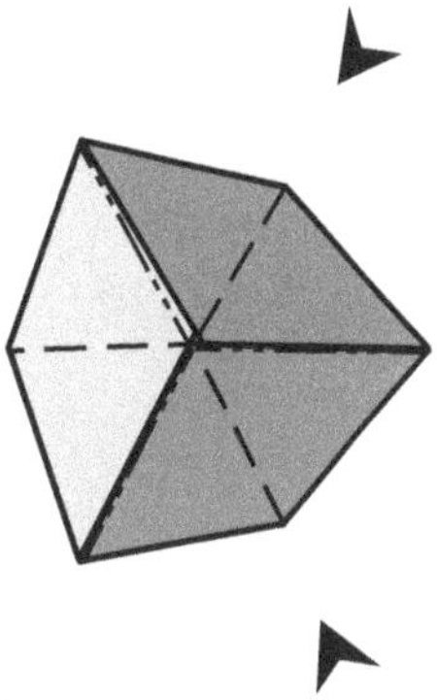

18 Continue squeezing into a 3D shape.

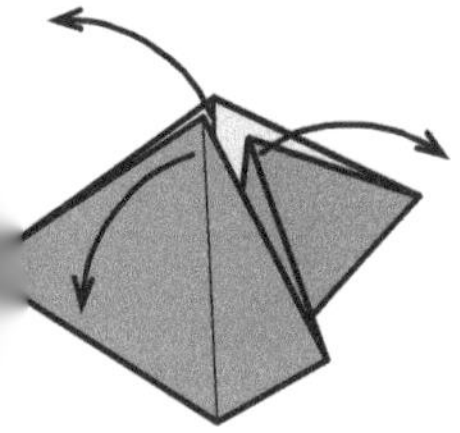

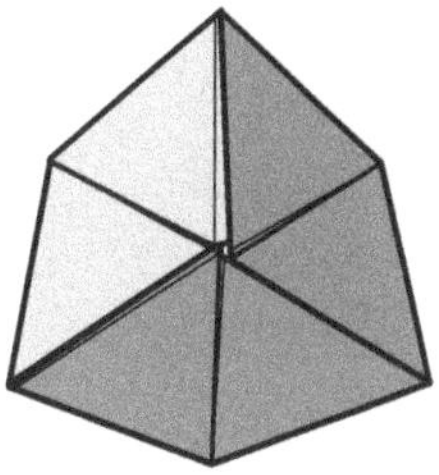

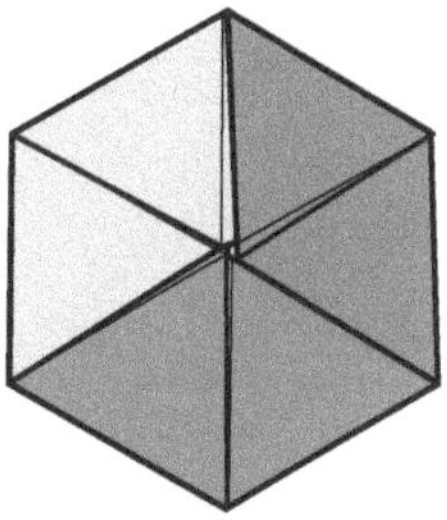

Open and flatten to reveal previously hidden faces: this procedure is known as the pinch flex.

ction

erform the pinch flex to reveal previously hidden faces.

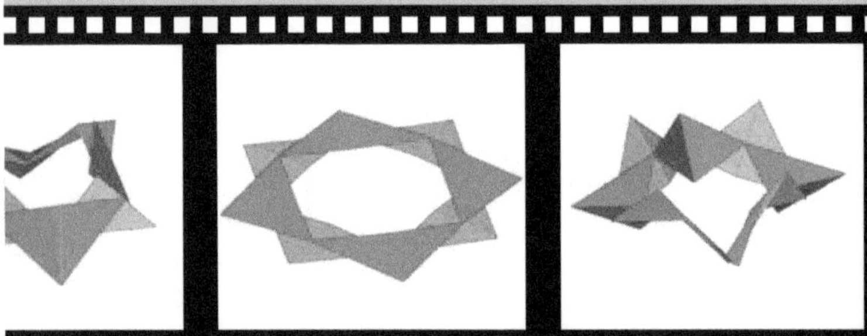

Rotating Octagram Ring

★★☆☆☆

Use 8 sheets of A

Finished model ratio 2.

I discovered this model whilst experimenting with waterwheels for this book of modular action origami. Some folders call this a "bonus model", a creation found whilst trying to make something else. The book fold in step 1 was only needed for locating other folds and would have been avoided in a later "clean" version. However, these folds were convenient for folding the model up so that it was smaller for transporting: they naturally make a preliminary fold or waterbomb base. I then found that not only did the model change between the preliminary fold and waterbomb base (like that folded from a square), but also turned completely inside out. This act of serendipity shows that a prepared mind can discover what it is seeking (modular action origami), even if it appears to be by accident.

Module ★★

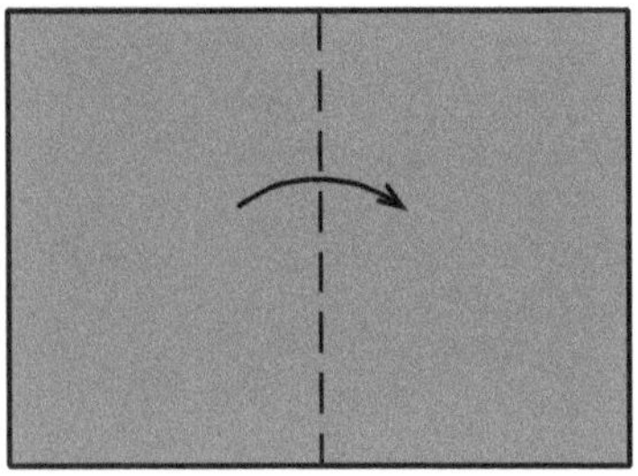

1 Fold in half.

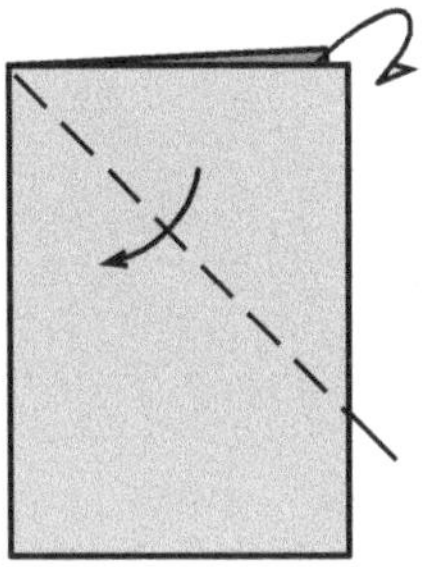

2 Fold the right corner down. Repeat behind.

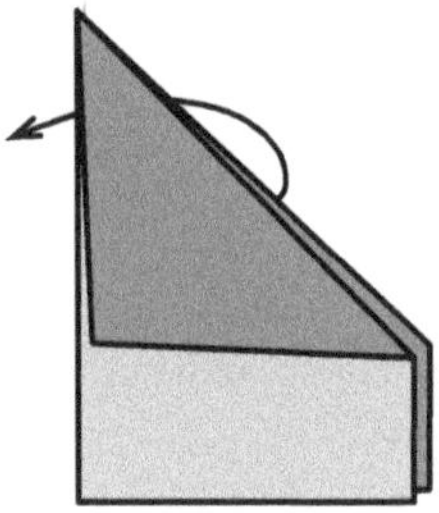

3 Unfold from behind.

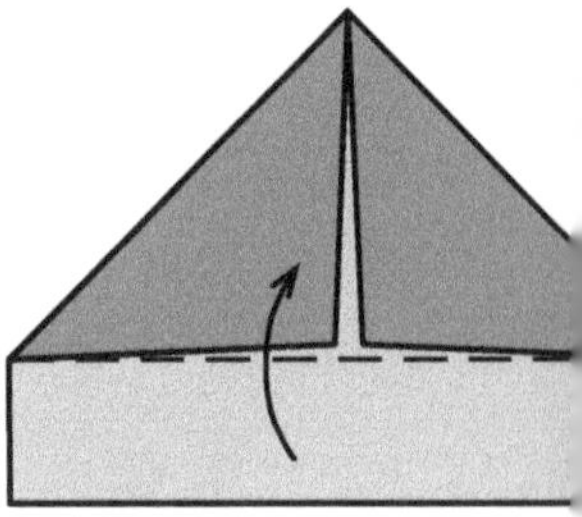

4 Fold the bottom edge up and ov the corners.

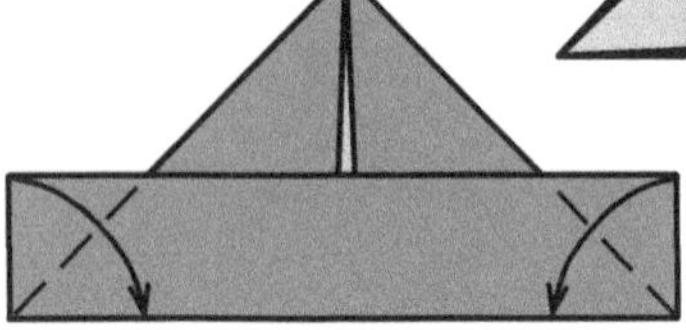

5 Fold the left and right corners down.

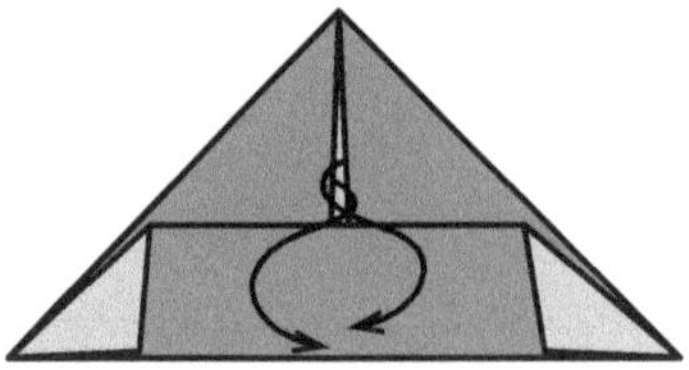

6 Tuck the whole flap inside the larger corners.

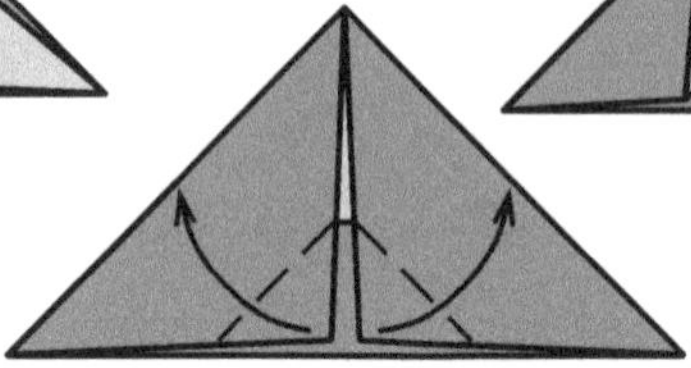

7 Fold the corners out using the hidden flap as a guide.

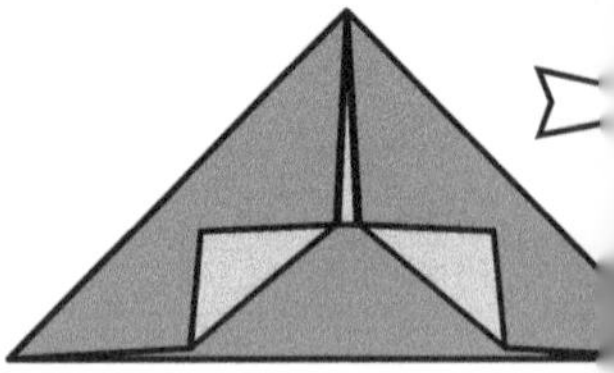

8 Unfold leaving t horizontal flap i place. For the second unit, full unfold.

Assembly ★★★

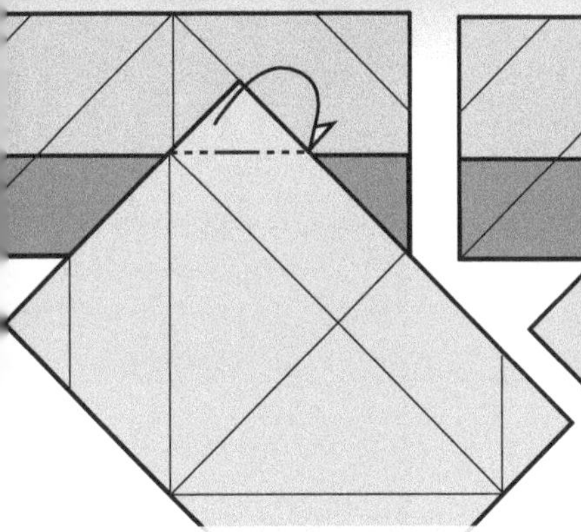

Tuck the top left corner of the second unit behind the flap of the first unit

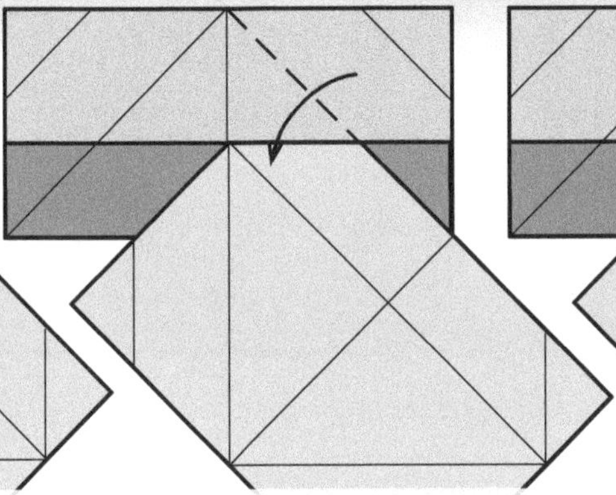

2 Fold the right corner of the first unit over the second unit.

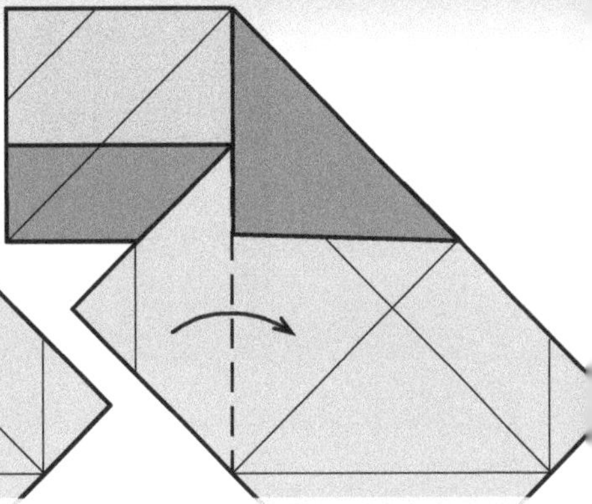

3 Fold the left corner of the second unit over the first unit.

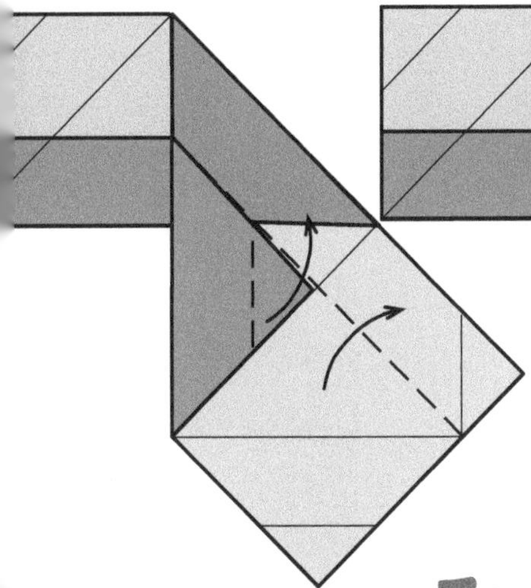

Fold the second unit over, tucking the corner into the pocket made by the first unit

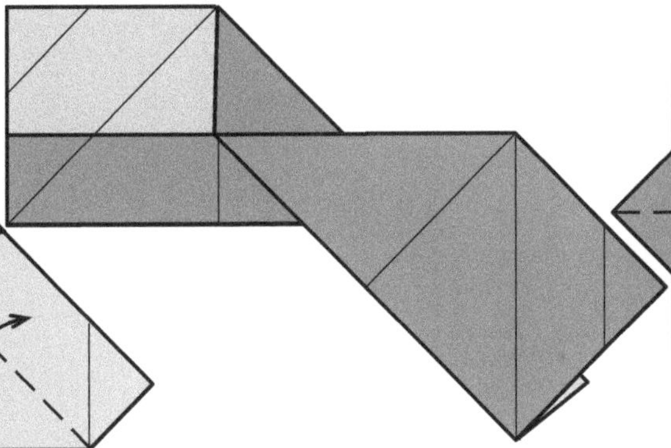

5 Sometimes you will need to turn the model over to add more units.

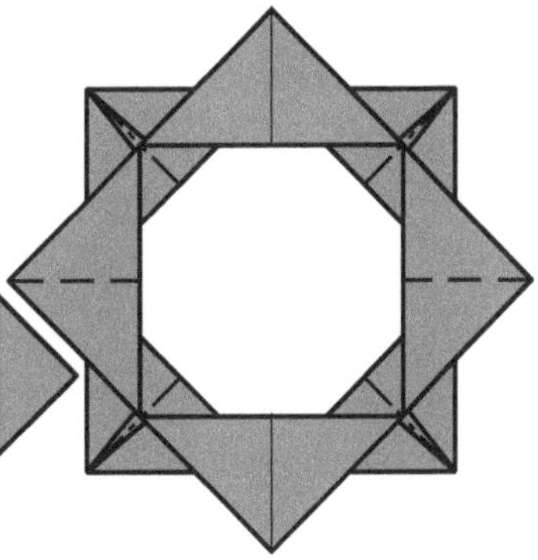

6 Reinforce the creases that form a waterbomb base.

Action

Make a waterbomb base. Turn the model inside out through the central hole.

12-piece Carousel

★★★☆☆

Use 8 squares about 10 cm
Finished model ratio 1.8

This is a 12-piece version of Carousel. Using more units makes the rotation smoother.

In module step 4, you can use Fujimoto's method of dividing an angle into equal parts. First fold the left edge with your best estimate and unfold. Fold the right edge to the crease you've just made. Unfold. Now fold the left edge to the crease just made - if your first estimate was good, this fold will coincide with your first estimate. If not, continue until the error becomes negligible.

Module ★★★

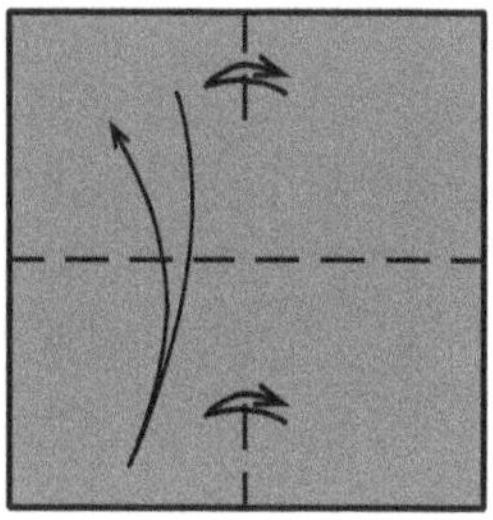

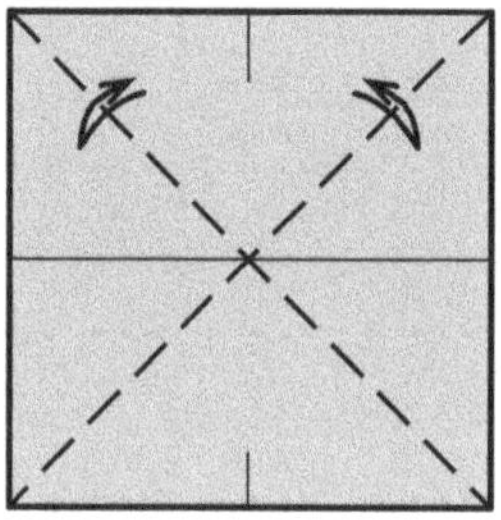

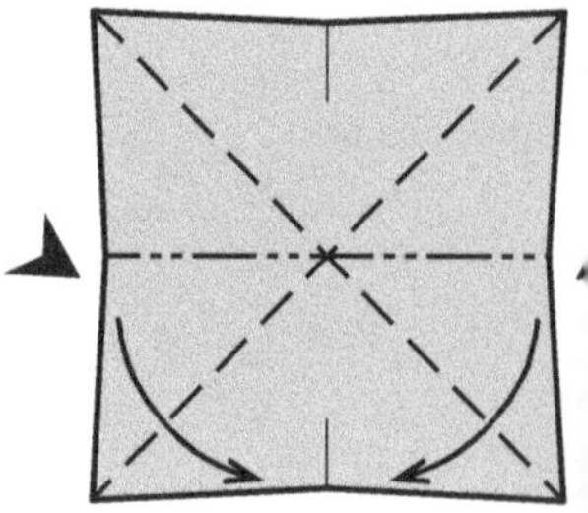

1 Fold the bottom edge to the top and unfold. Pinch the midpoints of the top and bottom edges. Turn over.

2 Fold and unfold each diagonal.

3 Use the existing creases make a waterbomb base Fold the top edge down and inside reverse fold th left and right sides.

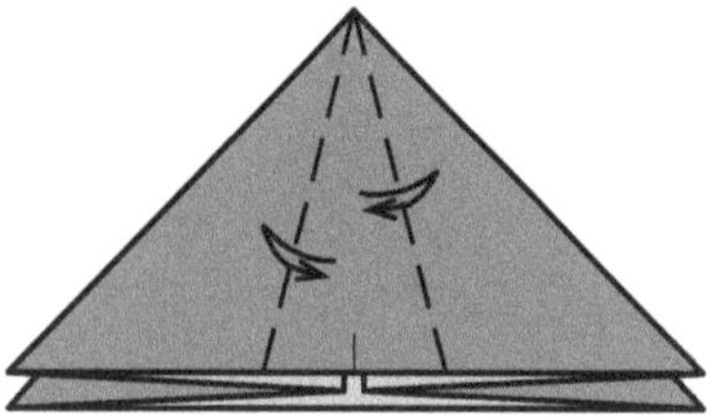

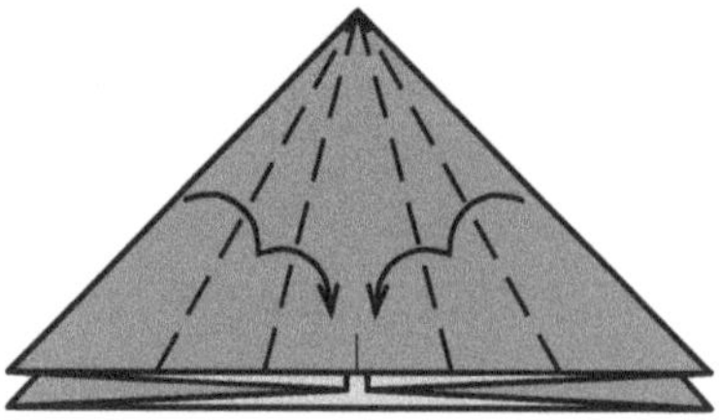

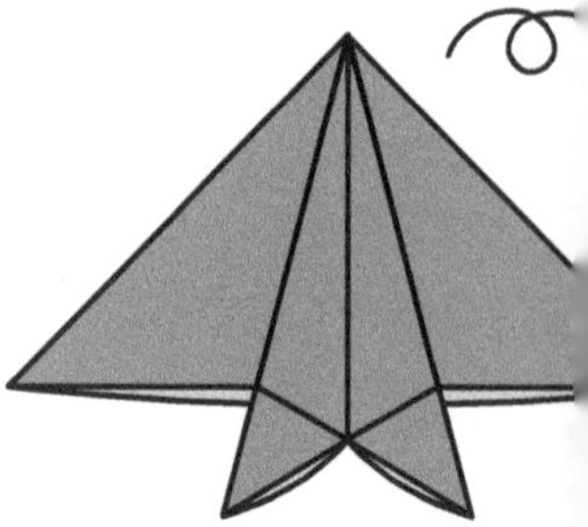

4 Divide the right angle into three equal angles. You can do this by eye with good accuracy.

5 Fold left and right flaps over and over to the centre.

6 Like this. Turn over.

Repeat step 5.

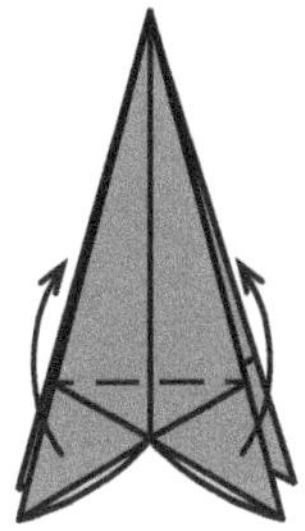

8 Fold all four flaps up. The fold lines align with the original edges of the square.

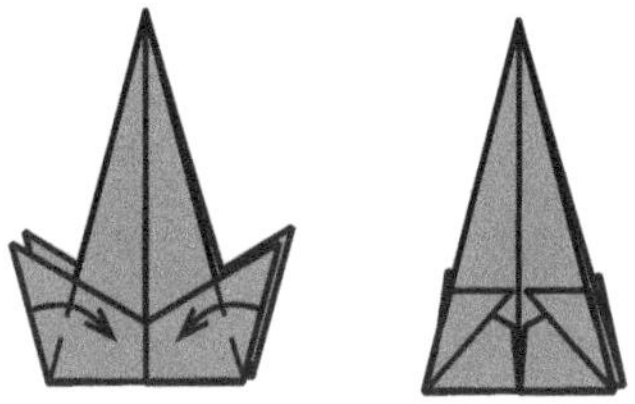

9 Fold each flap inwards using the folded layer behind as a guide. You need eight units in total. Two, three or four colours work well.

ssembly ★★★

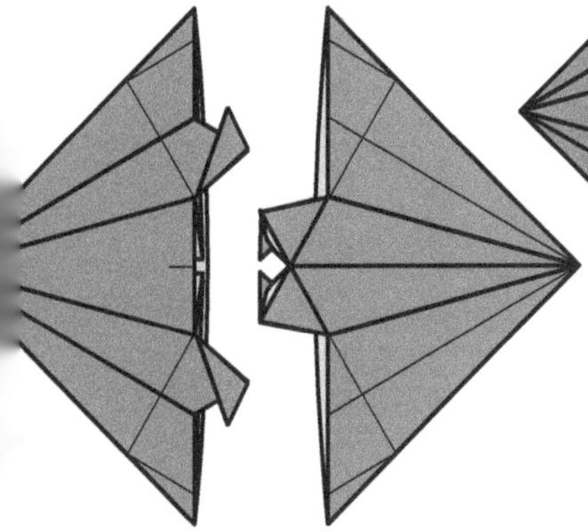

ake two units. To avoid ıe flaps interfering with ach other, unfold the rear ıps. Arrange the front ıps as show. Place the cond unit on top of the rst unit.

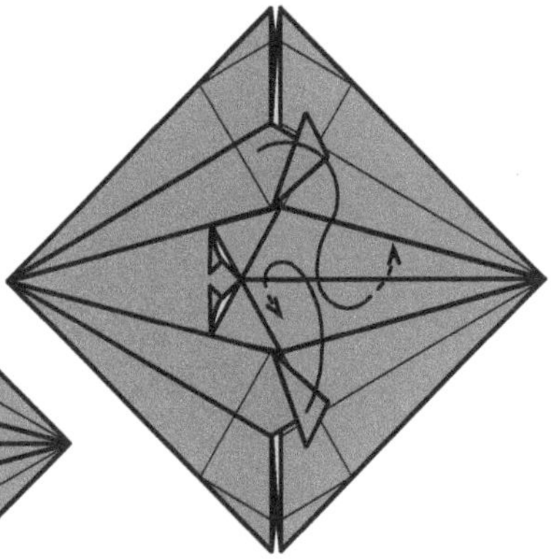

2 Wrap the flaps of the first unit around and inside the flaps of the second unit. The ends of the flaps tuck inside the second unit's pocket.

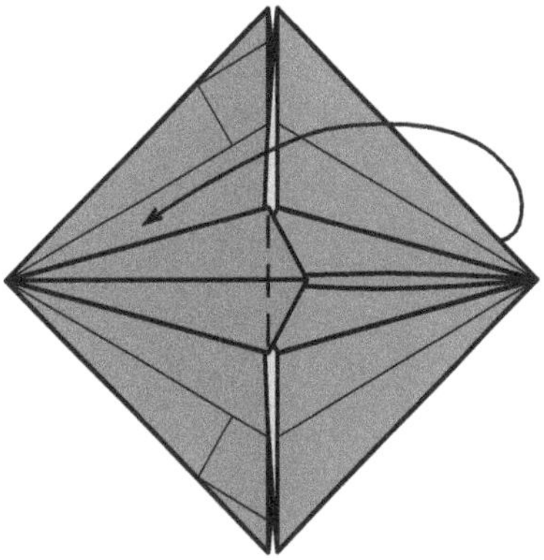

3 Valley fold the left unit onto the right unit. You can now add more units like in step 1.

ction

otate the points of the units around the circumference of the circle formed by the dges of the paper.

Multiform Flipper

★★★☆☆

Use 4 square

Finished model ratio 1.4

This model uses four 1:4 rectangles and shows different outline shapes when flexed. Make t A modules and two B modules. You can cut one 15cm square for this, or better still, use t squares so that you have more colour combinations.

You can also make this model from a single square creased into a 4 by 4 grid: slit the diagonal the central quarter square and alternately mountain and valley fold the triangles. Dave Mitche elegant *Ad Infinitum* combines four Pig Bases and has a fixed outline shape.

Module A ★★

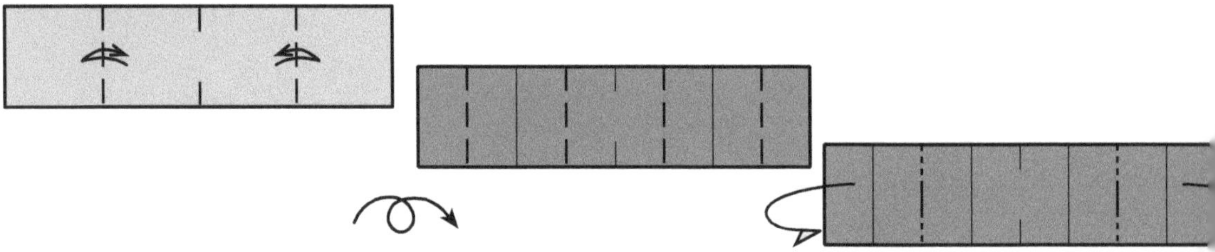

1 Pinch midpoints of the long edges then crease one and three quarters. Turn over.

2 Fold and unfold eighths.

3 Mountain fold quarters.

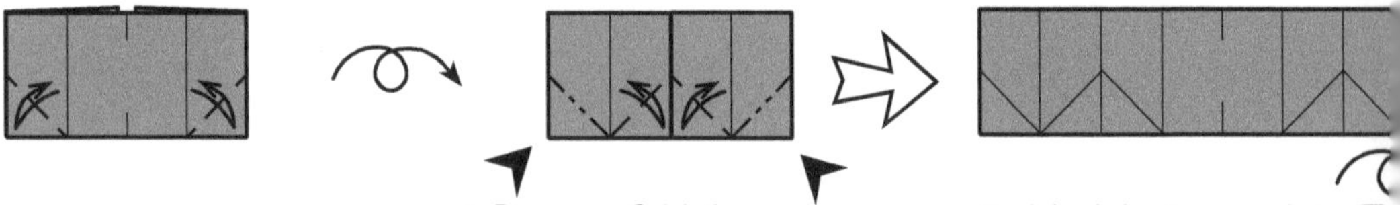

4 Fold and unfold the bottom corners. Turn over.

5 Reverse fold the corners. Crease the inner corners. Unfold completely.

6 Module A complete. T over ready for assembl

Module B ★

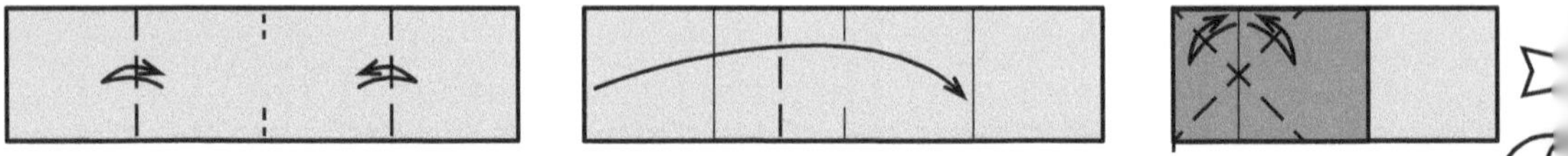

1 Pinch midpoints of the long edges then crease one and three quarters. Turn over.

2 Fold three eighths.

3 Crease the diagonals c left square, top layer o Unfold and repeat on right, then completely unfold. Turn over read assembly.

ssembly ★★★

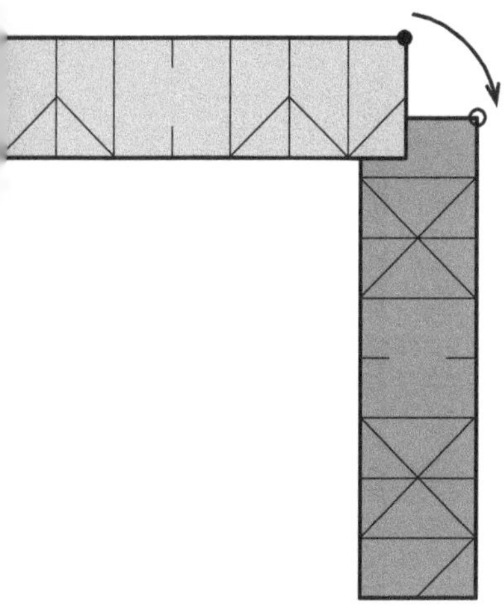

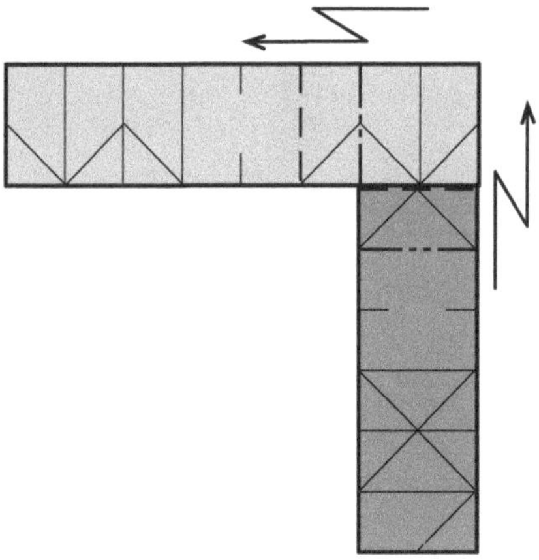

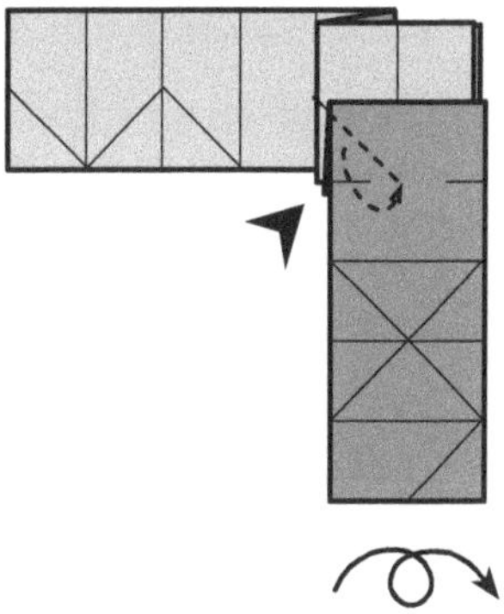

Put module A on top of module B. orientating the modules as shown.

2 Pleat the modules. Make sure that the layers are aligned.

3 Inside reverse fold the corner: the rear layer has one valley fold. Turn over.

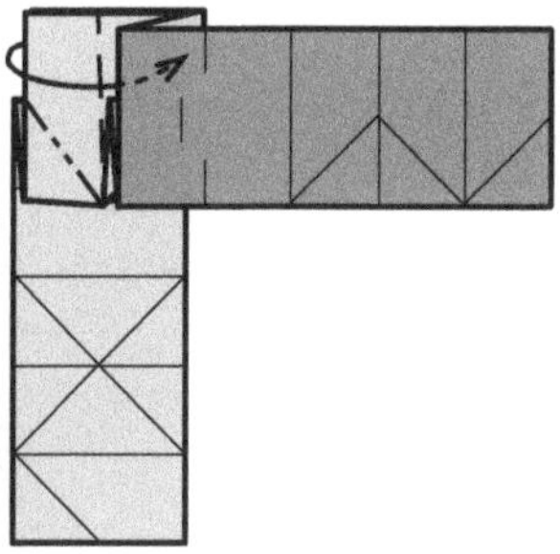

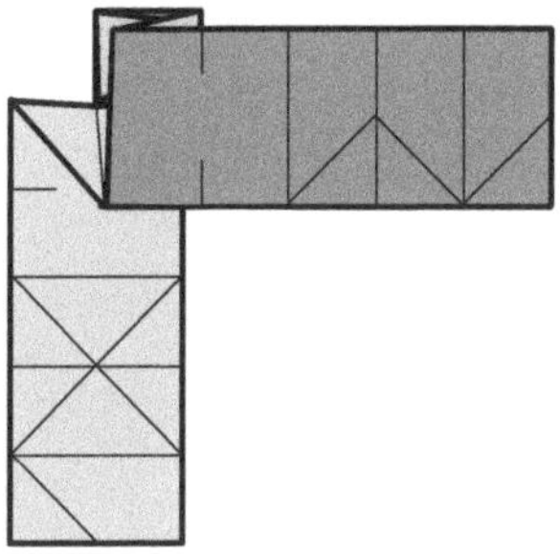

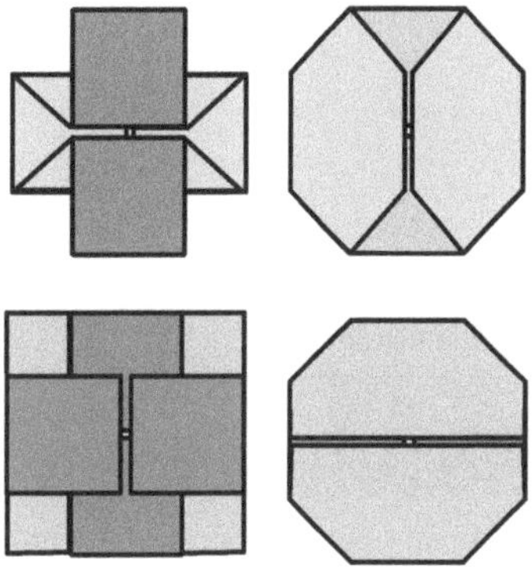

Squash fold. Tuck the flap inside the pocket, covering the reverse fold made in step 3.

5 Two units joined. Add the other modules using reflective symmetry,

6 These are the four positions obtained when you flip pairs of flaps behind.

Action

Flip the flaps behind, either one at a time or in pairs. The axis of rotation changes each time. If you flip pairs of flaps then you can make an irregular octagon, cross and square. What happens when you flip the flaps one at a time?

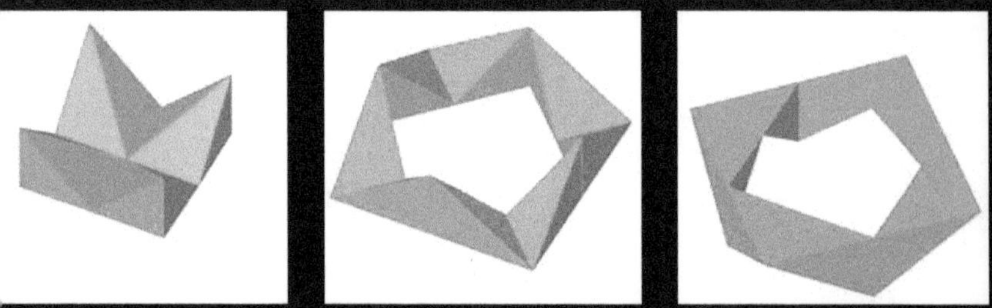

Double Skew Tetrahedra

★★★★☆

Use 5 or 10 squares
Finished model ratio 0.6

This is based on Thoki Yenn's Skew Tetrahedra [Yenn 87]. 15 cm origami paper is suitable. Ea unit comprises two tetrahedra joined by a hinge. You can connect four or six units using th joining tabs to make rings of eight or twelve tetrahedra. For the former use four squares for th units and one for the joining tabs; for the latter, use eight squares for the units and two for the joining tab.

Joining Tabs ★

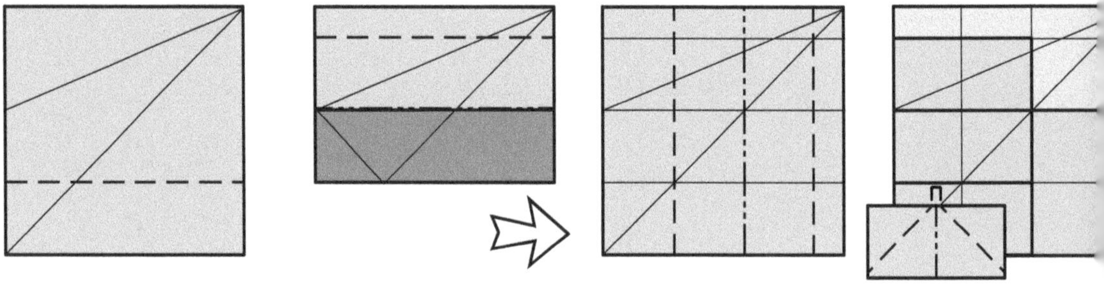

1 Precrease diagonal and angle bisector. Fold the bottom edge up.

2 Pleat the paper using the same height. Unfold.

3 Use the diagonal to make the same pleats vertically.

4 Cut four tabs fro the square. Crea the tabs with a middle gap 1-2m

Module ★★★★

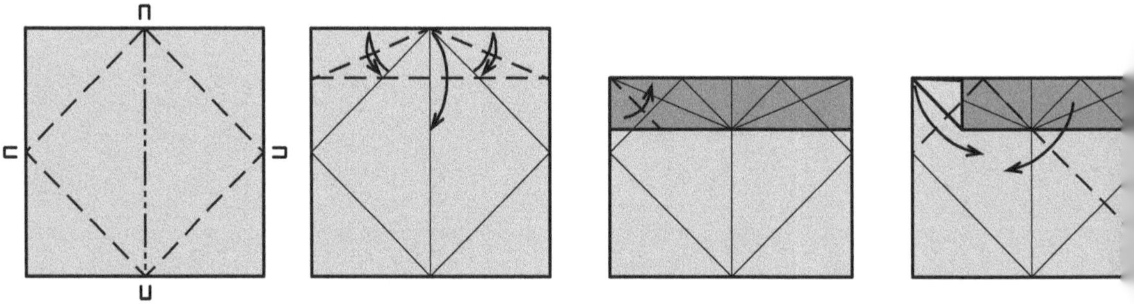

1 Precrease. Leave a tiny gap about 1mm.

2 Bisect then fold the top edge down through the ends of the creases.

3 Fold the left corner up.

4 Make two folds that extend the existing creases.

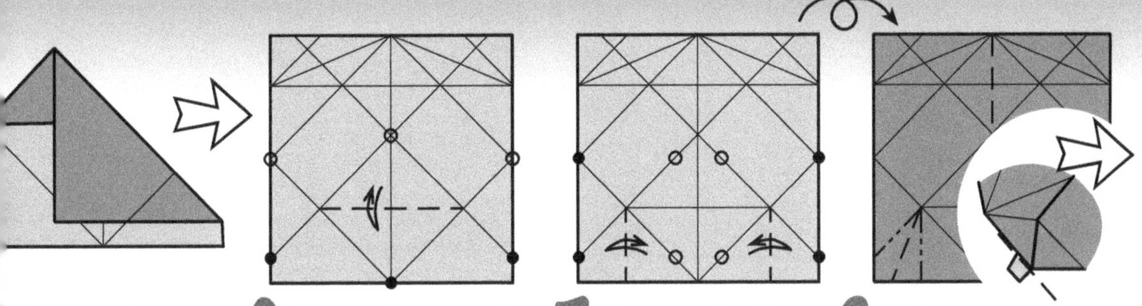

Unfold and repeat steps 2 to 4.

6 Crease the central square using the given reference points.

7 Crease using the given reference points. Turn over.

8 Make a convex cone and crease the protruding flap.

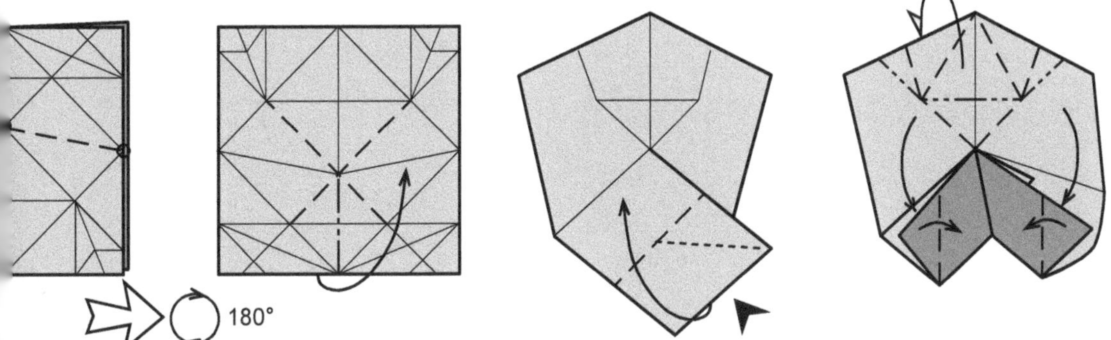

irmly crease hrough the two ocation points. Jnfold and turn.

10 Lift the centre and fold right to make a concave shape.

11 Fold the bottom edge left and squash the right flap.

12 Fold the corners from step 4 and then pull the top behind, closing the model.

embly ★★★

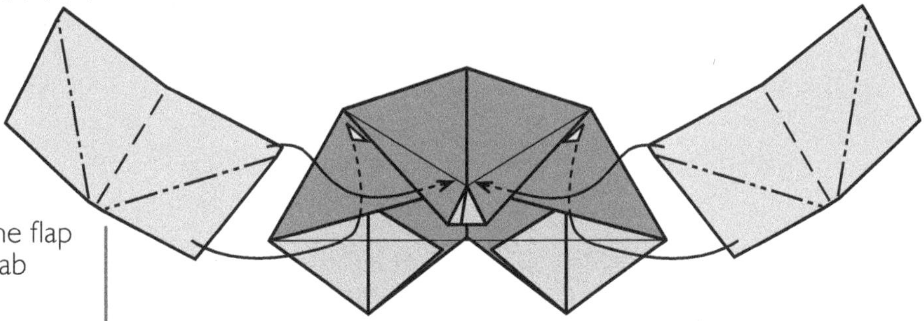

lide and tuck the flap f each joining tab nside the unit.

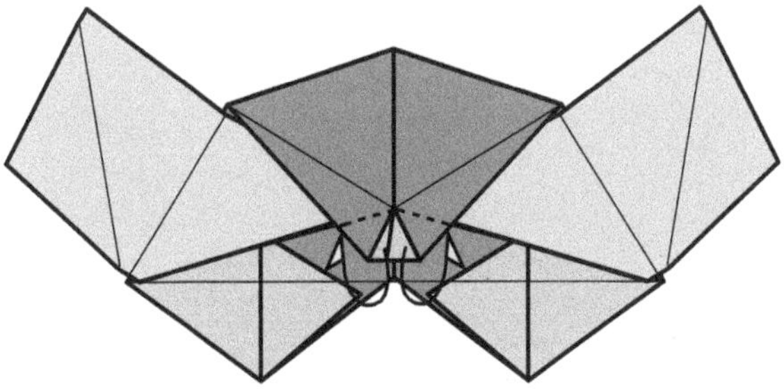

old the tabs behind nd inside: these tabs vere made in step 8. ecure the tabs by naking the bases of the wo pyramids lie flat.

3 Complete the unit by folding the flaps up and tucking the raw corners inside.

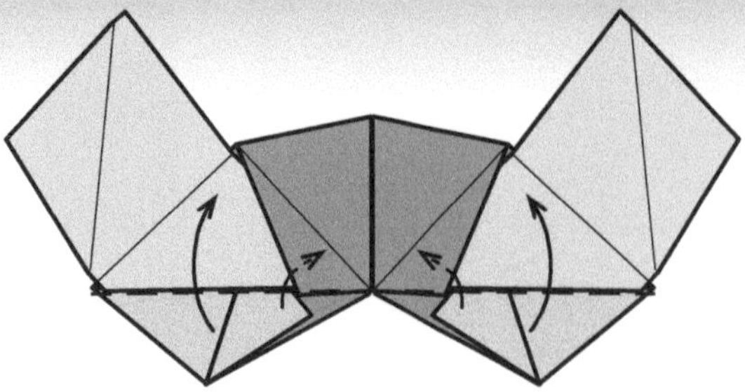

4 Add more units. Make a ring of four or six modules.

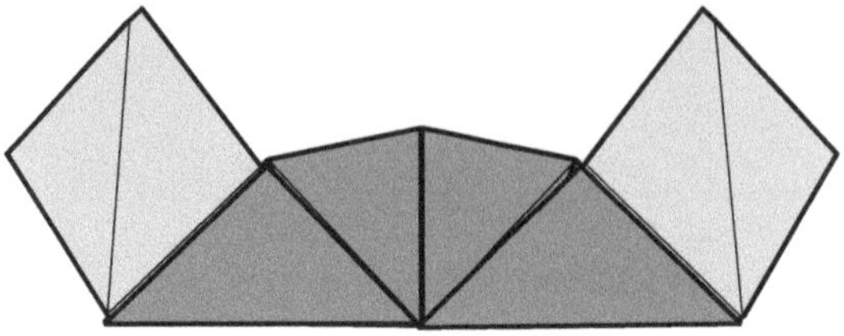

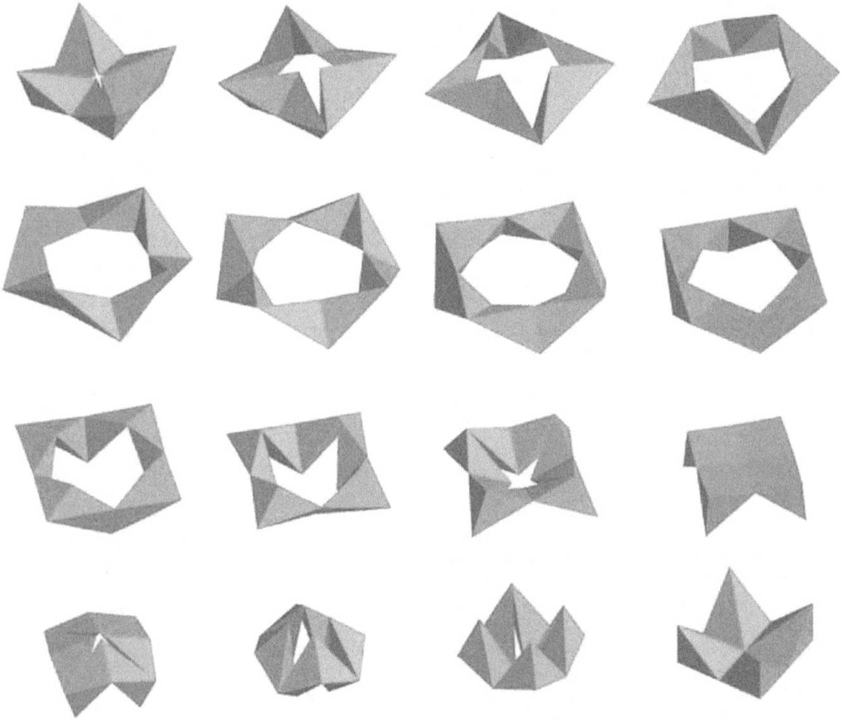

Action

Rotate the tetrahedra through the central hole. You can flex the twelve tetrahed version into many shapes.

Chapter 3

Magic Wallet Series

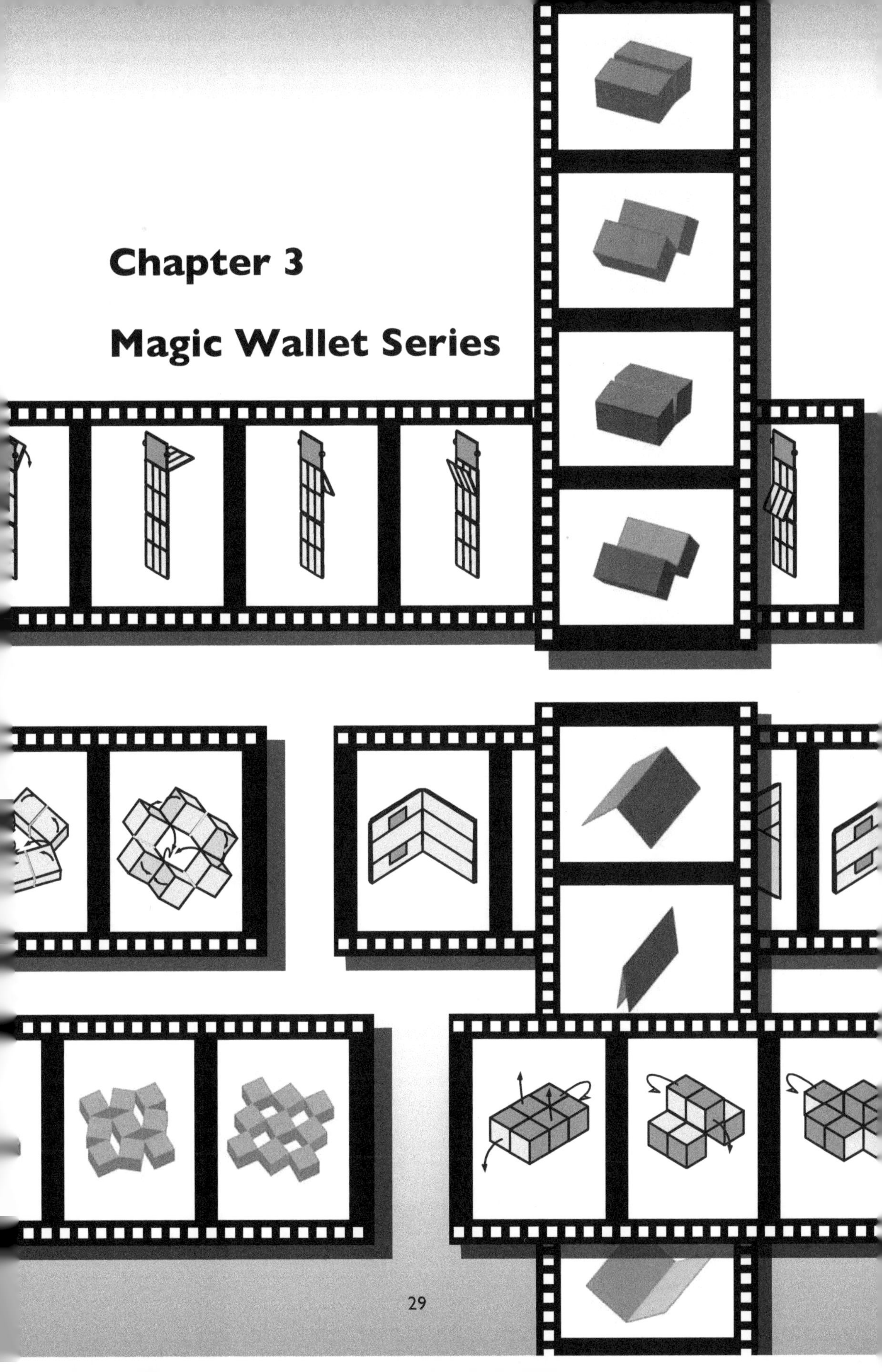

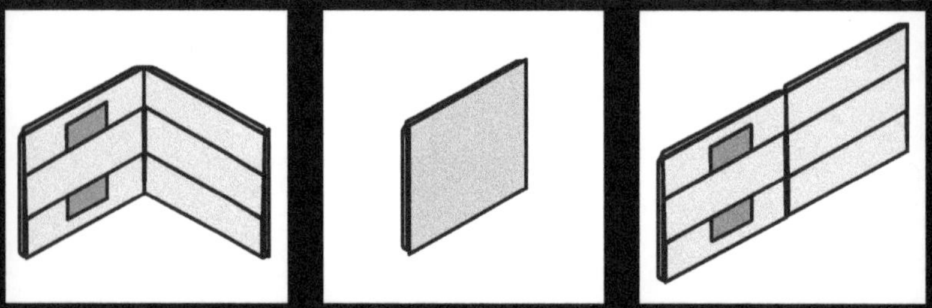

Magic Wallet
&Magic Wallet Series

★★★☆☆

Use 4 rail tickets and 1 sheet of A4 paper

Use rail tickets (85 x 54mm) and A4 paper (310 x 297mm). Or use cards about 1:1.4 to 1:6 and paper that is about 5.5 times the width of the card. The number of tickets that you need depends on what you are making. Each A4 rectangle makes 12 strips and four hinges.

Card Module ★

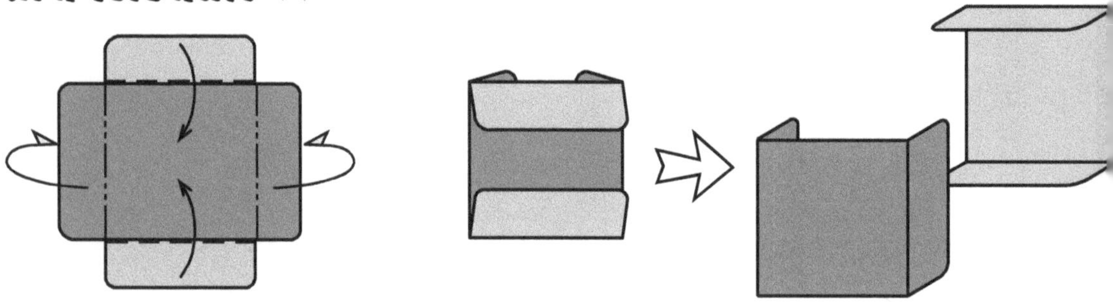

1 Place two tickets back to back. Fold the edges to form simple tabs.

2 Separate and keep the tabs at right angles.

3 Make many more units.

Joining strips ★

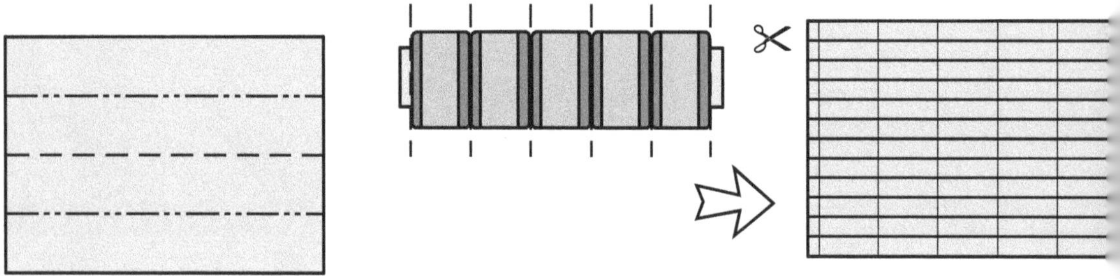

1 Fold into quarters.

2 Add creases separated by one card width plus 1 to 2 mm. Start the first card about 10mm from the edge of the A4 paper (about the width of the orange strip on a British rail ticket).

3 Divide into twelfths and cut into strips.

aking the Magic Wallet Hinge ★★

To make the Magic Wallet Hinge, put two units edge to edge. Take three strips and weave them between the two units, starting with the bottom strip. The strips are tightly wound cept between the two cards: leave a gap of about 2 to 4mm. (The gap needed depends on hich model you are making: most models need a gap of about 1 to 2mm except Jacob's Ladder at needs a gap of about 5mm.) Add the middle strip and then finish with the top strip.

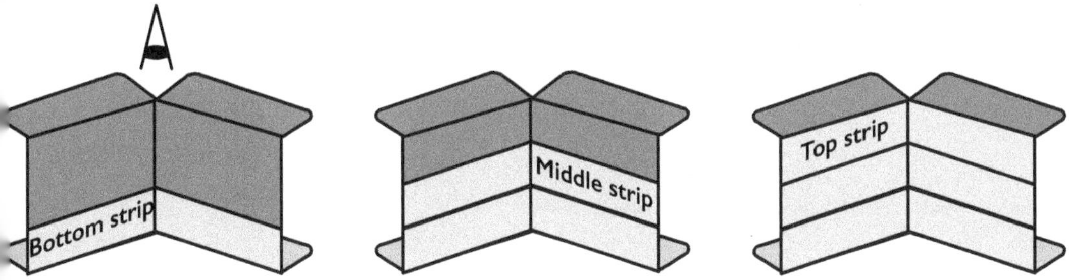

Top strip

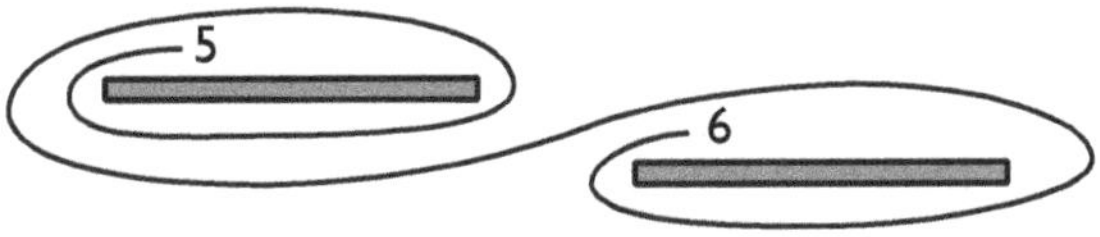

5 Put the third strip here and wrap strip around the left ticket...

6 ... then wrap around the right ticket, finishing here.

Middle strip

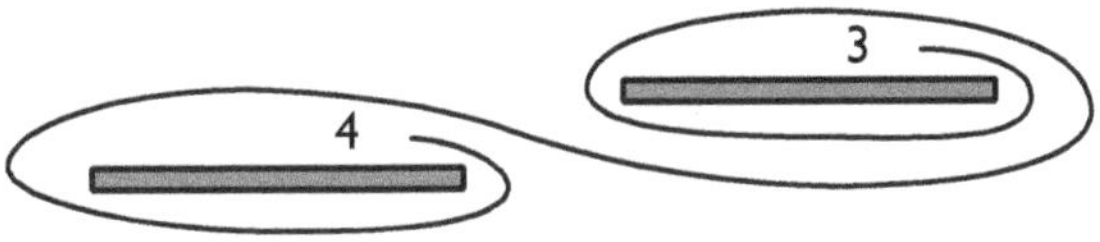

3 Start wrapping the second strip here on the right ticket...

4 ... then wrap around the left ticket, finishing here.

Bottom strip

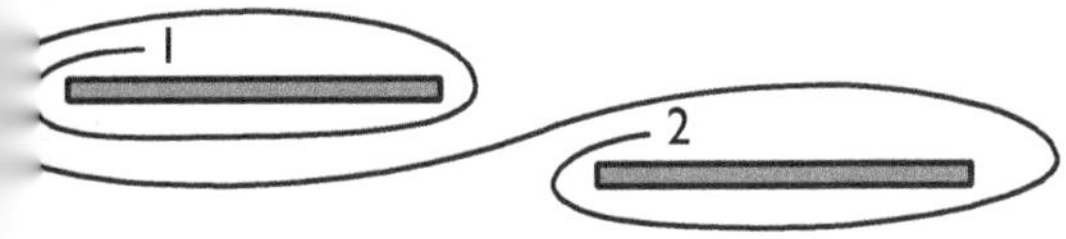

Put the first strip here and start wrapping around the left ticket...

2 ... then wrap around the right ticket, finishing here.

Assembly ★

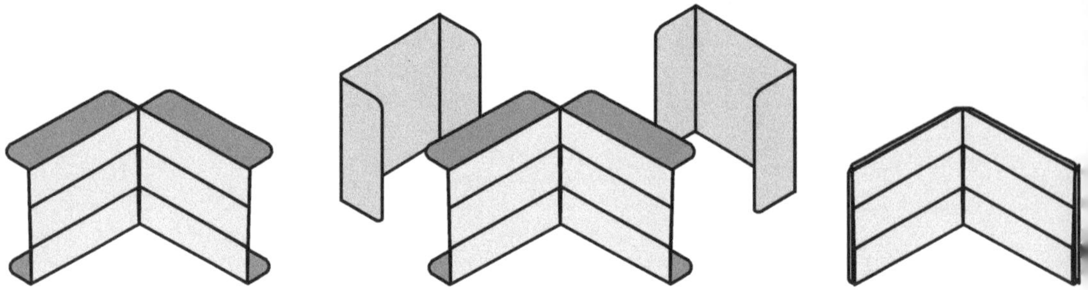

1 Make a magic wallet hinge.

2 Take two more cards and interlock each with each half of the hinge.

3 Magic Wallet complet

How to use the magic wallet to change a red card into a blue car

1 Prepare the wallet by inserting the blue card into the right side behind all strips.

2 Show the audience the red card and insert it into the left side so that is held by the middle strip.

3 Close the right side o wallet and the open t left side. The red card appears to have beco the blue card.

Action

You can vary the effect to make an audience member's banknote disappear or to change value. Folded are easier to interweave with the wallet's strips. Depending on the orientation of the wallet and its stri can make the note appear to be behind the middle strip or behind the top and bottom strips.

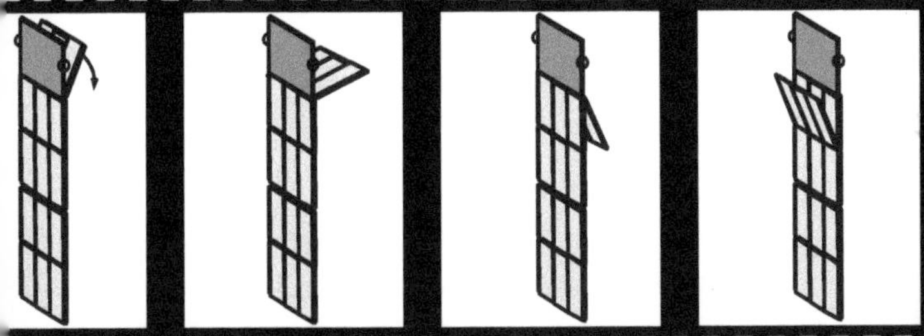

Jacob's Ladder

★★★★☆

Use 16 rail tickets and 1 sheet of A4 paper

or this model you can use narrower strips than usual. Divide the A4 sheet into sixteenths instead of twelfths. Leave a gap of about 5mm between the two cards of each hinge: if the ɔ is too small then the joints are too tight and the model will not work properly.

e ladder consists of three rows: the front row has three hinges, the middle row five cards and back row has two hinges. Join the hinges using the cards in the middle row. In the diagram n together the cards and units showing the same number. Numbers 3 to 6 have three cards t 1 and 2 only have two.

he cascade stops partway or is erratic, add weight to the units by inserting one or two cards ide, each folded in half. This should improve the action, but don't add too much weight or the der will swing too much.

he action is still erratic, you can add momentum to the first motion. For example, you could k the second card in the two-handed method. Or you could tilt the first card back more.

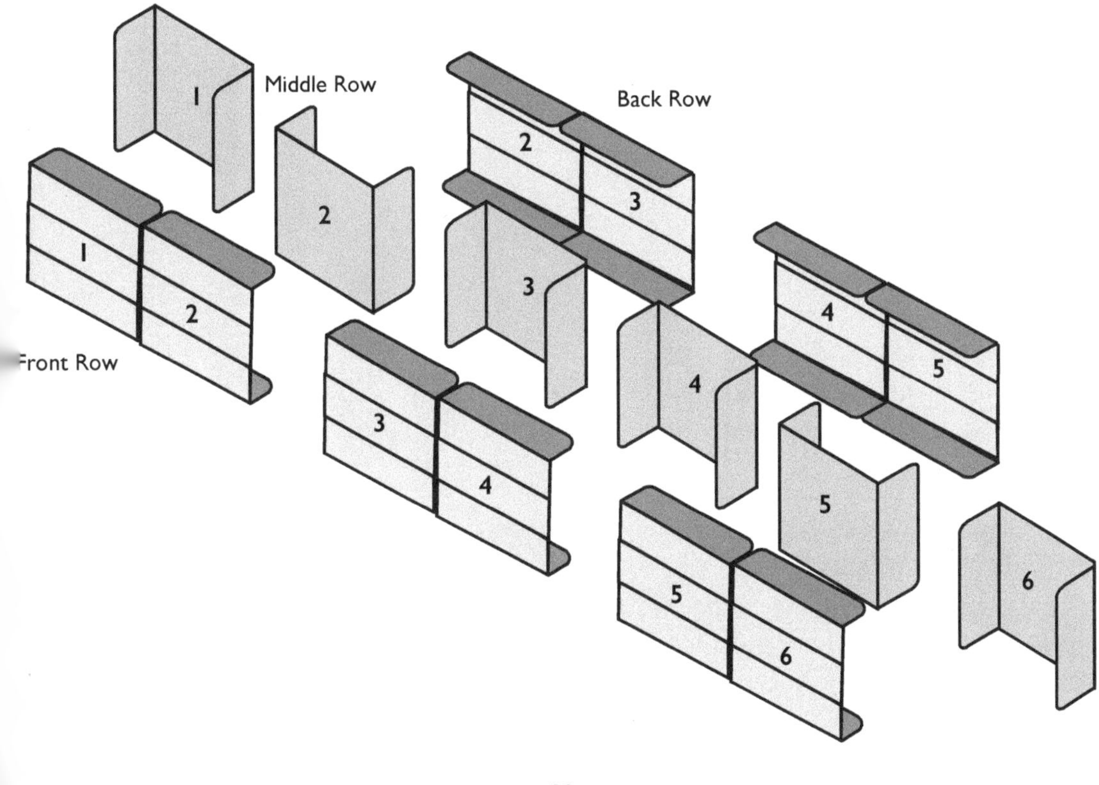

Here are two ways of working the ladder.

Two hands: Hold the second unit of the ladder. With your other hand fold down the top unit. Hold the top unit and let go of the second unit. This makes the lower units flip in turn.

One hand: hold the top unit and and turn it about a horizontal axis to send cascades downwards. Rotate the other way to bring your hand back to the first position. You don't have to wait for the cascade to finish before sending another cascade down. Repeat as often and as fast you like.

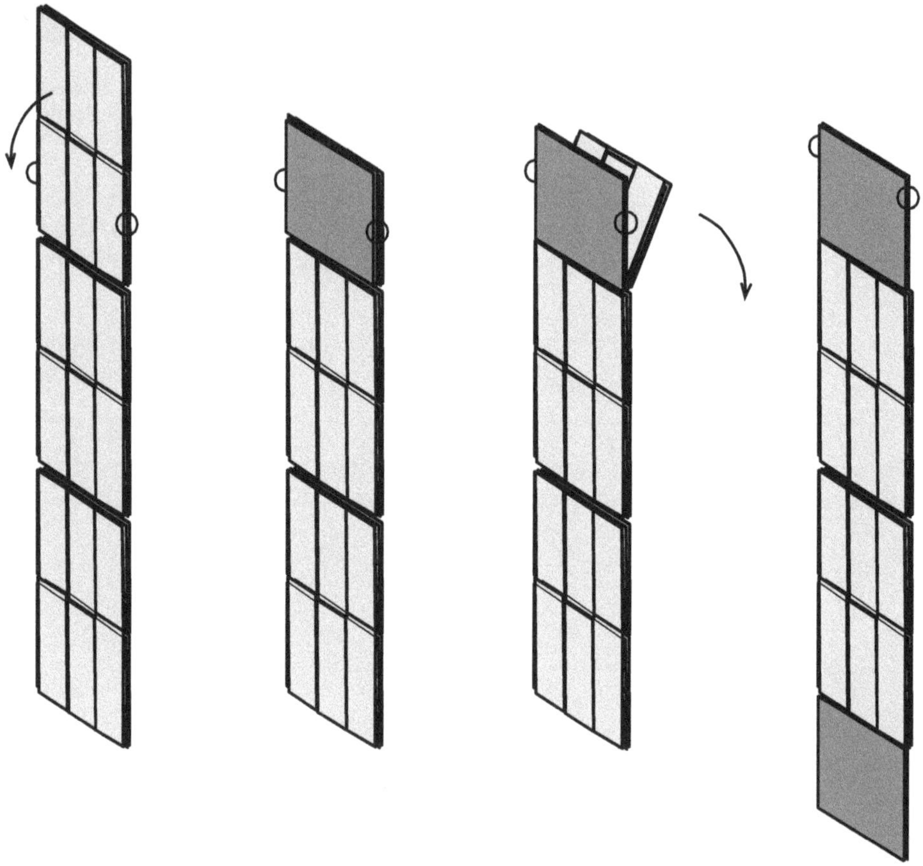

Action

You can work the ladder with one or two hands. Try the two-handed method befor the one-handed method. Unlike most other models, you can store this model folded in a repeated Z shape: this can help keep the creases sharp and improve the action.

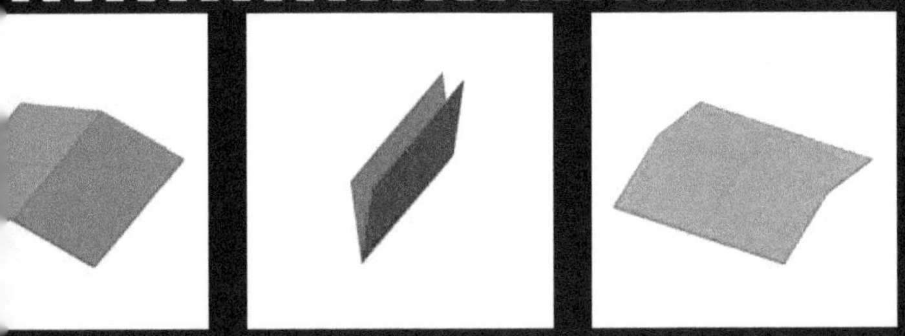

Four Square Flexagon

★★☆☆☆

Use 8 rail tickets and 1 sheet of A4 paper

ake four magic wallet hinges. Connect them in a chain so that the hinges are on the outside of the model. The hinges are perpendicular to each other.

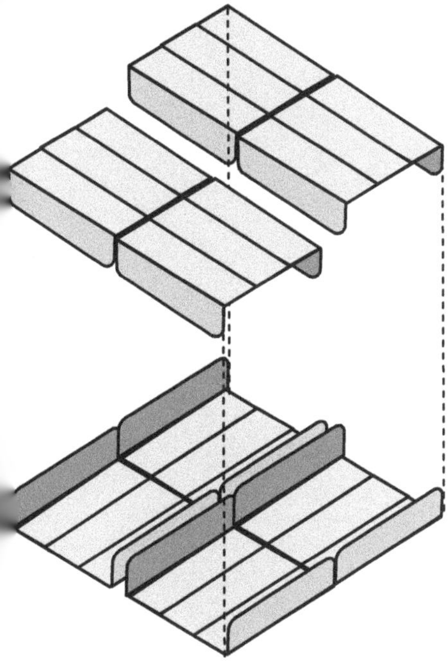

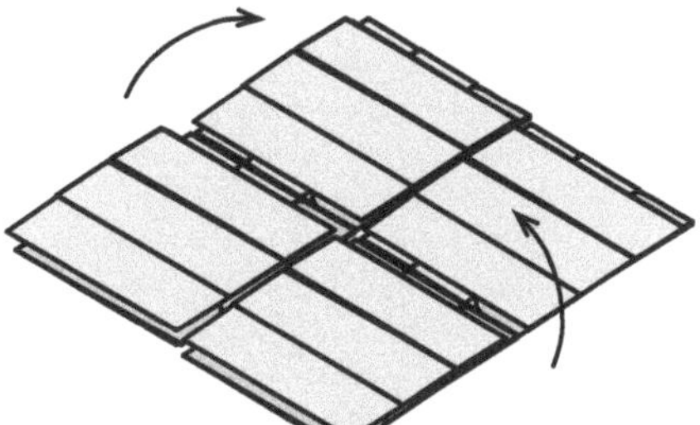

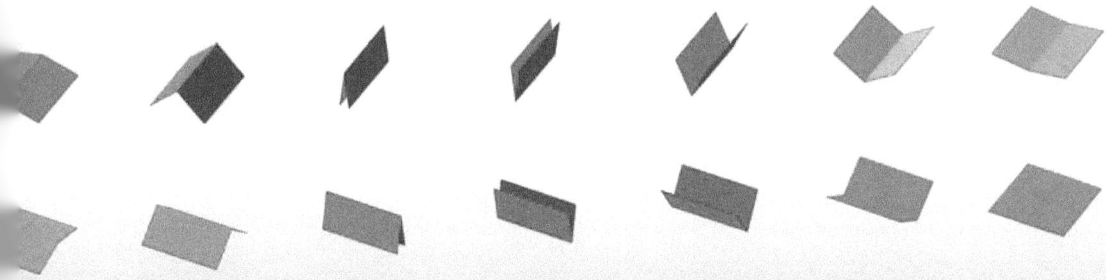

ction

om the first position, close the model like a book. Now imagine that the closed model is a book with spine uppermost: open the book from below. You can now repeat the moves but at right angles. This ngs you back to the first position so that you can repeat the moves as many times as you like.

Four Cube Flexagon

★★★☆☆

Use 52 rail tickets and 1 sheet of A4 paper

Make four cubes and clad five faces of each cube. Arrange the cubes into a 2 by 2 grid as shown below: the unclad faces are on the top of the right and back cubes and on the bottom of the left and front cubes. Arrange the cubes so that they will accept four magic wallet hinges. The connections are equivalent to the Four Square Flexagon.

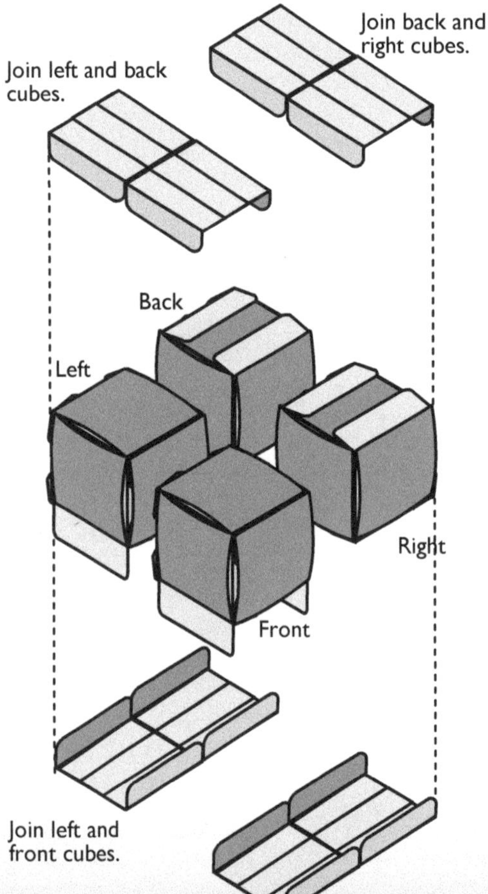

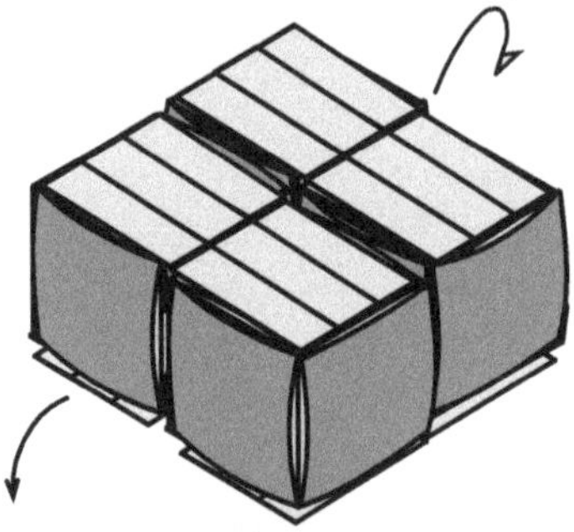

Action

Flex the model along the magic wallet hinges. You can flex twice in one direction and then flex twice in the other direction. This brings you back to the first position from which you can repeat the moves as many times as you like.

Making a Cube ★★

To make the Magic Wallet Hinge, put two units edge to edge. Take three strips and weave them between the two units, starting with the bottom strip. The strips are tightly wound except between the two cards: leave a gap of about 2 to 4mm. (The gap needed depends on which model you are making: most models need a gap of about 1 to 2mm except Jacob's Ladder that needs a gap of about 5mm.) Add the middle strip and then finish with the top strip.

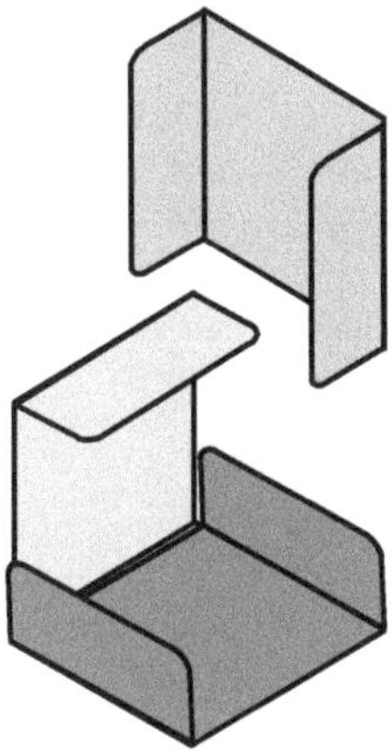

Position the units so that the tabs will be on the outside of the cube.

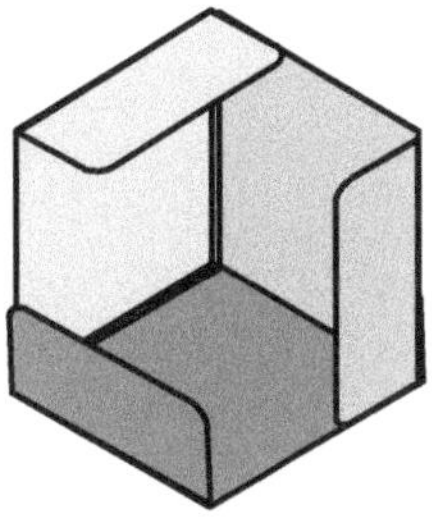

2 Three units positioned.

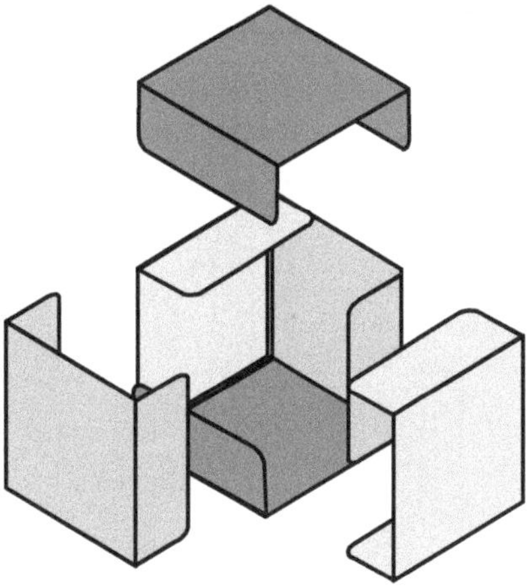

3 Slide the other units into place.

Six units joined.

5 You can secure the cube by cladding with extra units. The amount of cladding needed depends on the model.

6 Usually at least one pair of opposite faces needs cladding. Here the top and bottom faces have been clad.

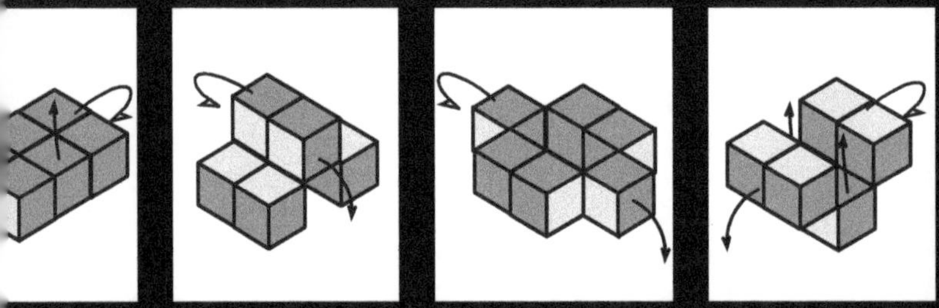

Six Cube Flexagon (Pseudo-Fluxicube)

★★☆☆☆

Use 80 rail tickets and 2 sheets of A4 paper

Make six cubes and eight hinges. You also need four cards to clad each cube plus one extra for the front two and back two. This means you need a total of 80 cards and two sheets of A4. This model is effectively the four cube flexagon in the second position with a pair of cubes inserted into the middle.

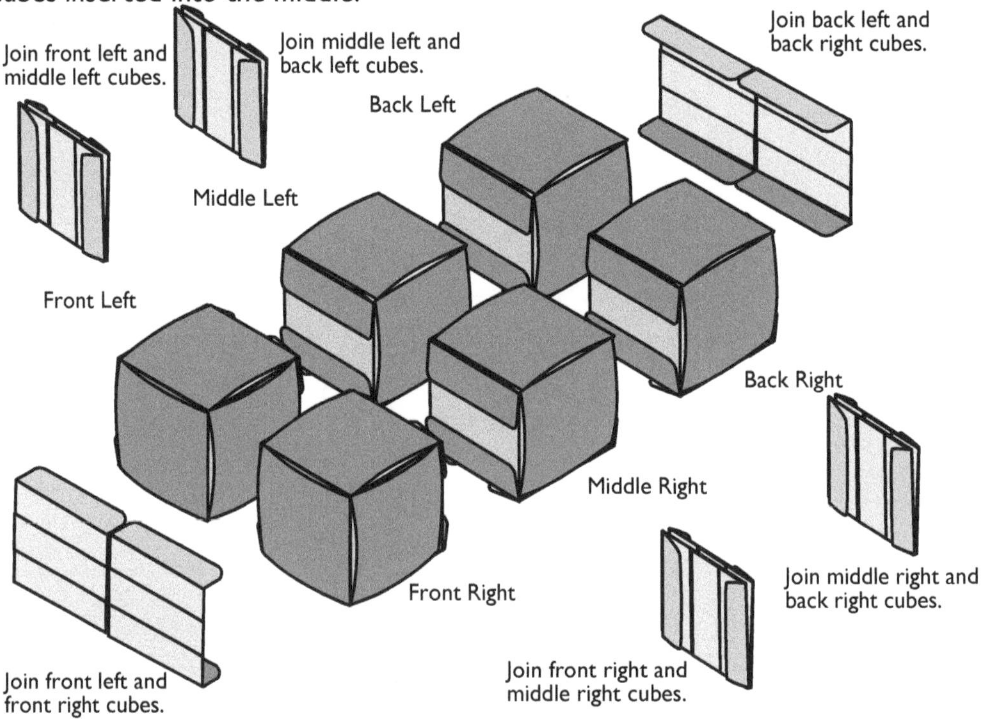

1 Put six cubes in two rows of three. Arrange the cubes so that the units of the top faces have short sides next to each other.

2 Clad the left, top, right and bottom faces of each cube. In addition, clad the front faces of the front cubes and the back faces of the back faces.

3 Use two hinges in each of cubes to a make a ch of three cubes. Join the of the two chains with t more hinges; these hing are at right angles to th hinges in the chains.

Action

Lift up the middle cubes and the front and back cubes will fall down. Separate the two middle cubes and rotate the away from each other. The front and back cubes will also rotate. Continue rotating the middle cubes until they mee again. Lift up the middle cubes and the front and back cubes will fall down, returning you to the first position.

Fluxicube

★★★★☆

Use 84 rail tickets and 1 sheet of A4 paper

his is a six cube flexagon but a different orientation of the hinges creates a very different flexing action. You need six cubes, six hinges and six cladding cards per cube. The gap veen the two halves of the hinges can be tighter than normal i.e. a gap of 1 or 2mm is ugh. Too wide a gap gives a loose and floppy motion.

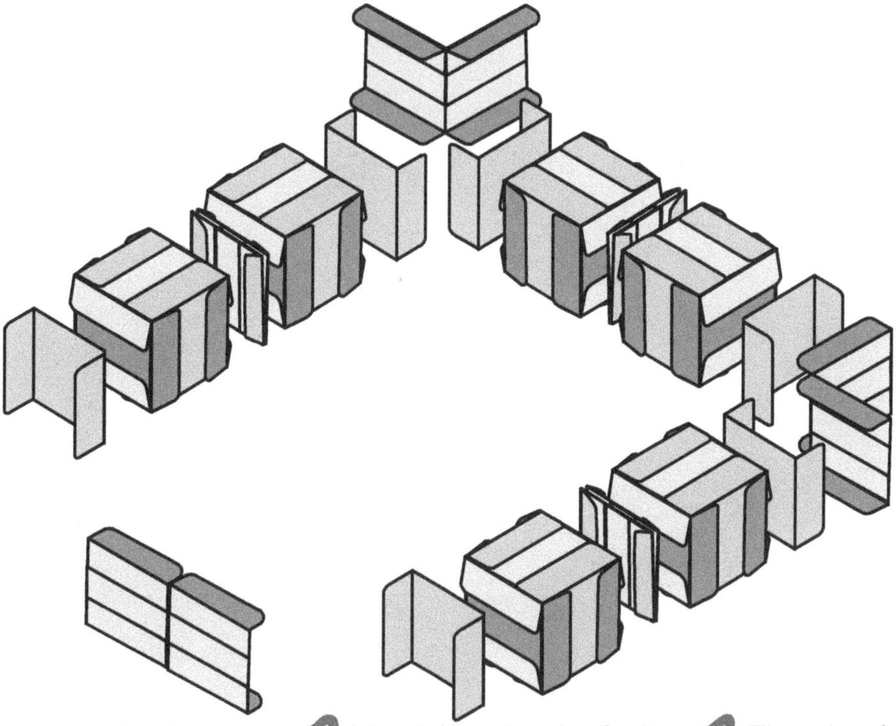

Arrange six cubes into three groups of two as shown. Join the two cubes in each pair with a hinge that flexes horizontally.

2 Now join each pair of cubes with a hinge that flexes vertically. You will need to clad each of the end faces to do this. This cladding may have the tabs inside or outside the cube.

3 The cubes form an equilateral triangle. Clad the remaining faces i.e. the top, bottom and faces next to the hinged faces.

ction

u can rotate the cubes in your hands, but their weight and size means a supporting surface is useful. t the model on a table and turn one cube: the other cubes will move in response. The motion is rprisingly smooth, and the model remains stable at most stages of motion.

Magic Wallet Hinge
Wobbling Wall

★★★☆☆

Use 9 cubes an
12 magic wallet hinge

Imagine a square lattice with alternate squares removed. If the the vertices of the squares are hinged then the lattice can be transformed; each square hole becomes a rhombus. This wobbling wall uses magic hinges so that you can transform it further than a standard wobbling wall.

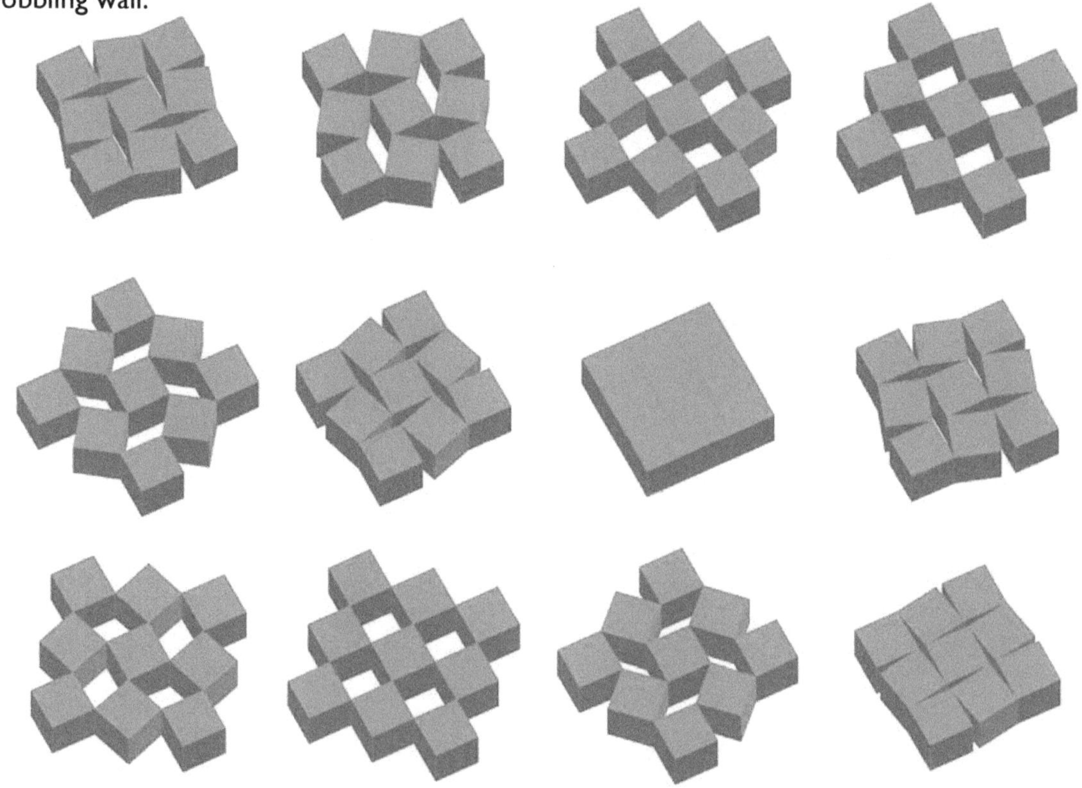

1 Arrange the nine cubes in a 3 x 3 grid. Use 12 magic wallet hinges to connect the faces.

2 The faces of some cubes have their tabs in the wrong place: clad the cube's face to "rotate" the tabs 90°.

3 When joined, clad the remaining faces i.e. the one or two other face not joined by hinges.

Action

Grasp a cube in each hand and turn the cubes. The magic wallet hinge allows you to rotate a cube around another cube from one edge to the opposite edge.

Chapter 4

Spinners and Wheels

Three-piece Square Spinner

★★☆☆☆

Use 3 squares
Finished model ratio 0.9

This spinner is inspired by a Makato Yamaguchi spinner. My version avoids petal folds but do not spin quite as well. However, it does feature a double square octagram that I have been slightly obsessed with. 75mm origami paper works well.

A and B Modules ★★

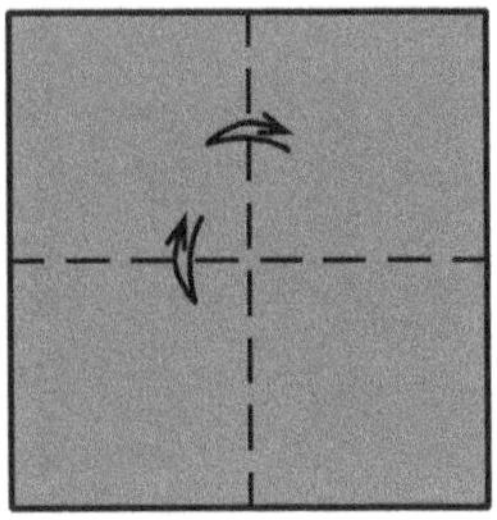

1 Crease a 2 by 2 grid. Turn over.

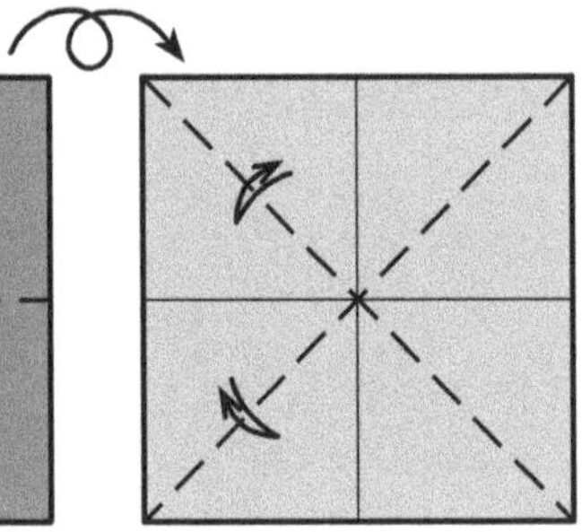

2 Crease the diagonals.

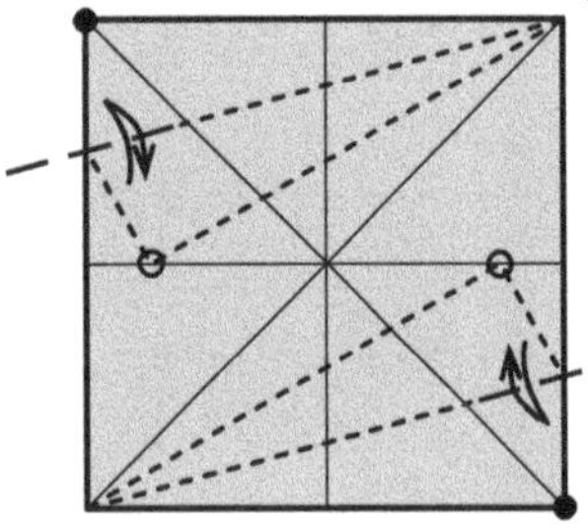

3 Pinch the edges.

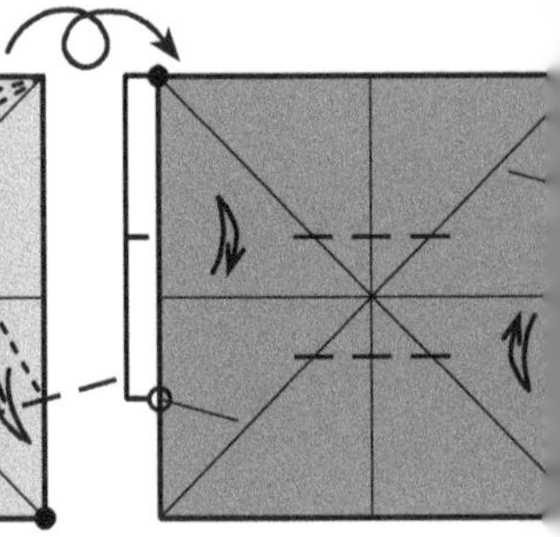

4 Pinch the diagon

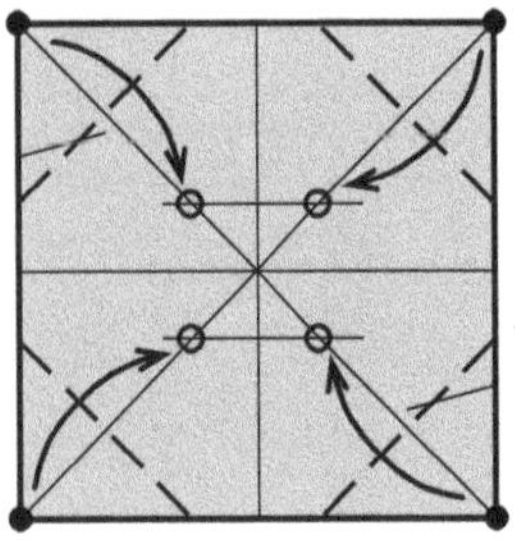

5 Fold the corners to the location points made in step 4.

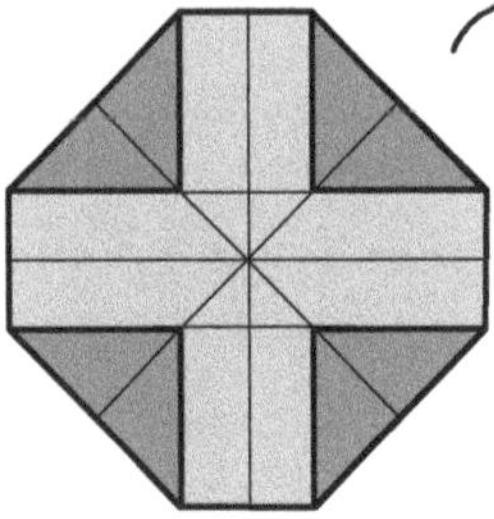

6 Turn over.

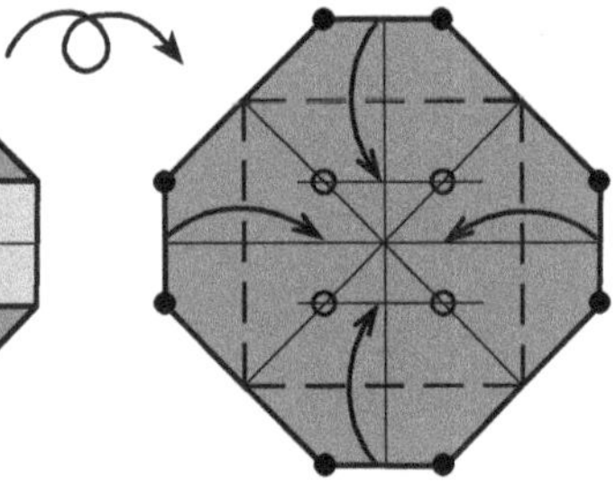

7 Fold the sides in.

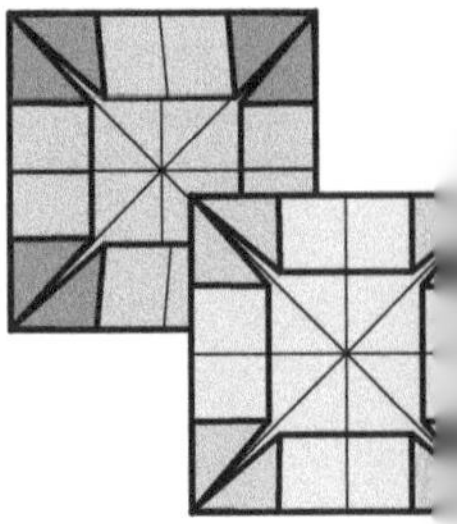

8 Make two of the modules.

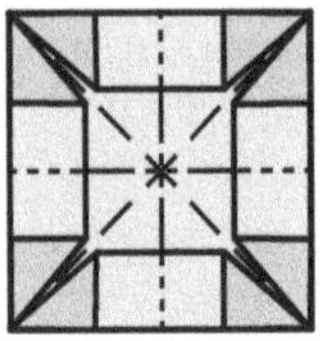

9 Module A: fold a waterbomb base.

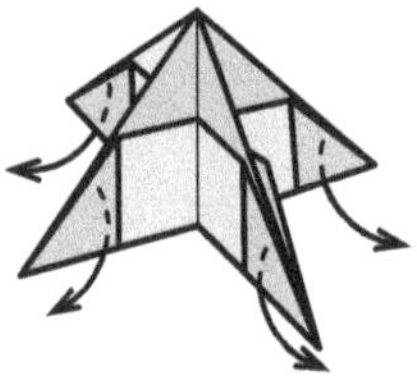

10 Open module A and pull out the corners.

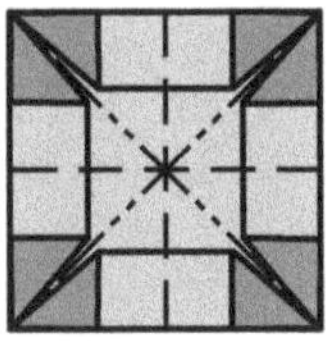

11 Module B: make a preliminary fold.

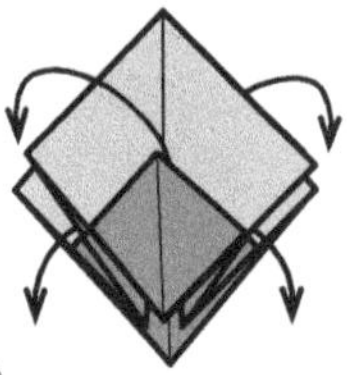

12 Open module E and pull out the corners.

Module ★★

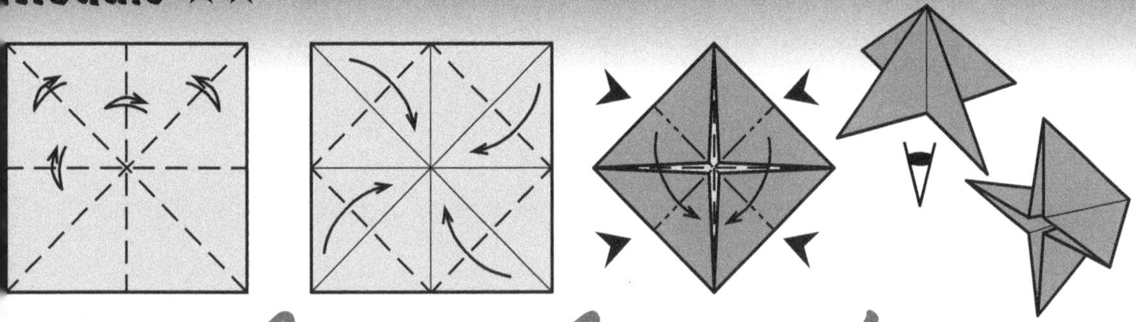

Crease in half.

2 Fold all corners to the centre.

3 Form a waterbomb base.

4 Here are two views of the completed Module C.

sembly ★★

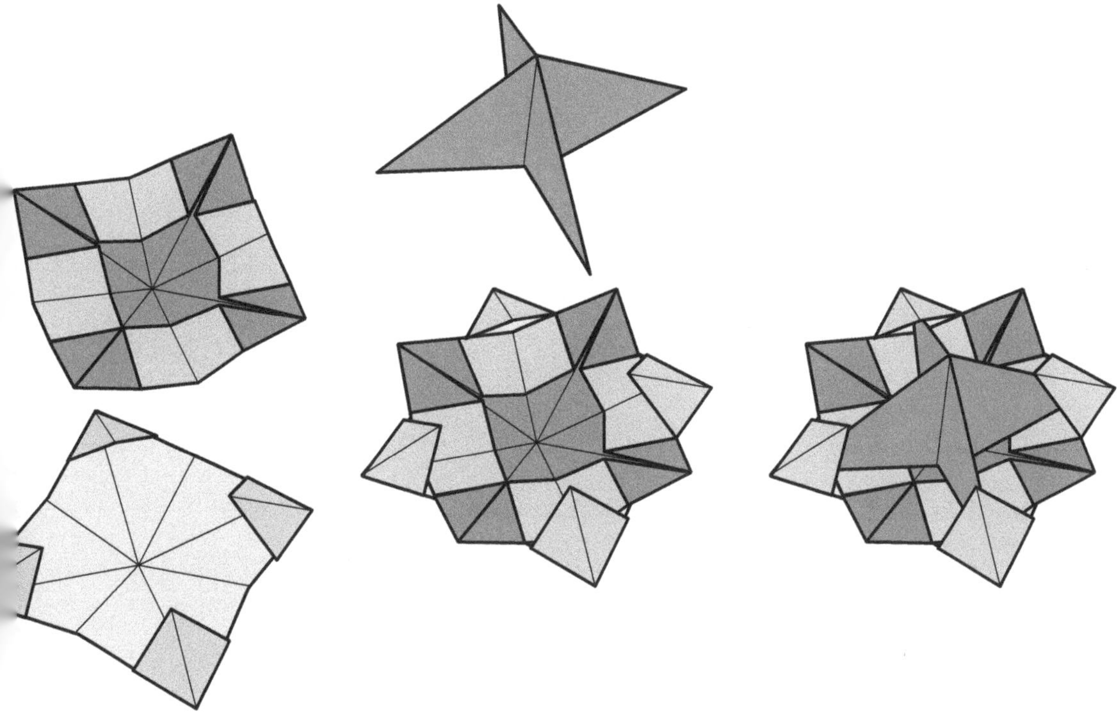

nterlock the corners of A and B modules.

2 Put the corners of C module the the corner pockets of B module.

3 Complete.

ction

ace the spinner on a flat smooth surface. Hold module C, spin and let go.

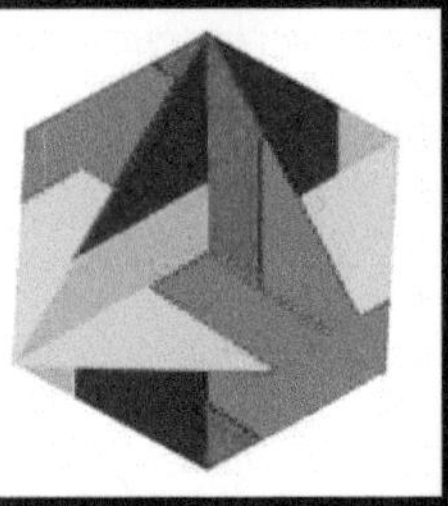

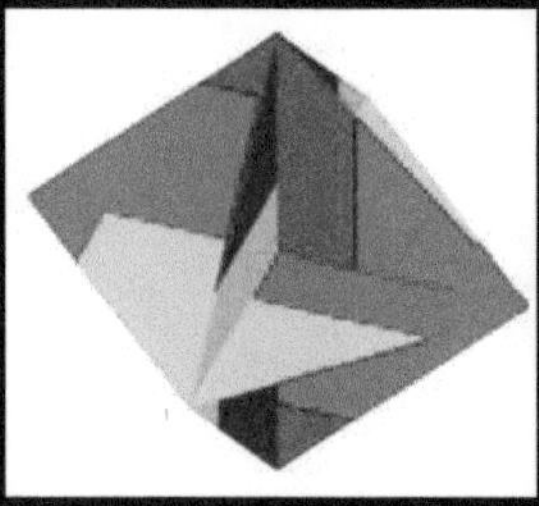

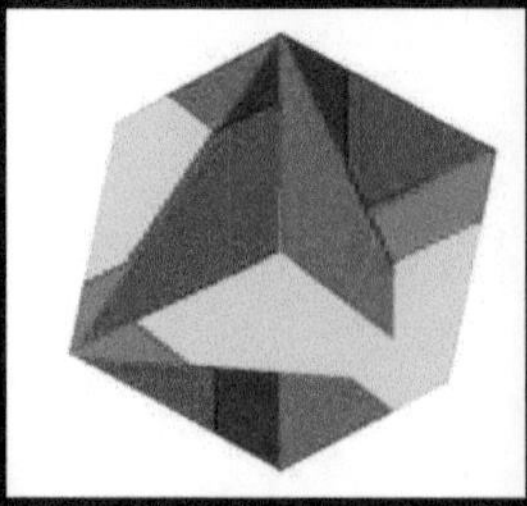

Skeleta Octahedron

★★☆☆☆

Use 6 sheets of A

Finished model ratio 1.

There are many origami skeletal octahedra. This skeletal octahedron uses a similar assemb pattern to Robert Neale's classic *Skeletal Octahedron*. However, using the partial side pock makes the assembly easier as you don't need to unfold the units during assembly. However, th vertices are less sharp and the assembly is slightly less secure.

This shape can be interpreted in many ways. It is three intersecting squares. It is the skeleton regular octahedron because the outer edges can be filled with eight equilateral triangles. It ca be fitted into a cubic lattice where the cube's vertices meet.

I first learned how to spin Robert Neale's classic *Skeletal Octahedron* from a book by Kunihik Kasahara.

Module ★

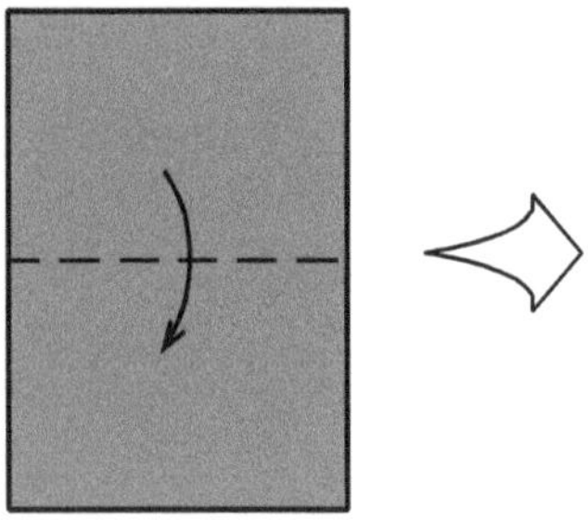

1 Fold in half.

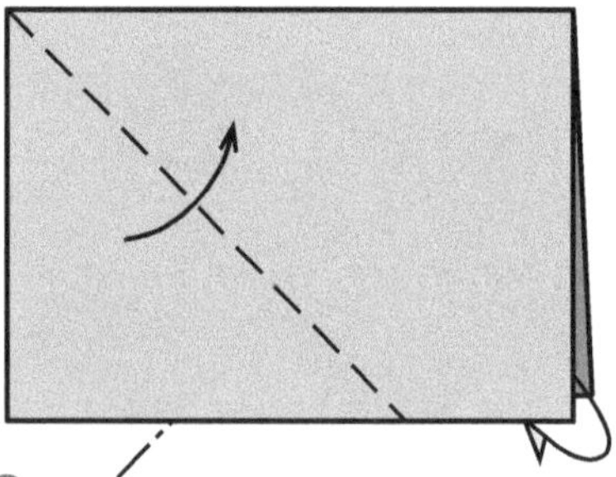

2 Fold the left corner up. Turn over and repeat behind.

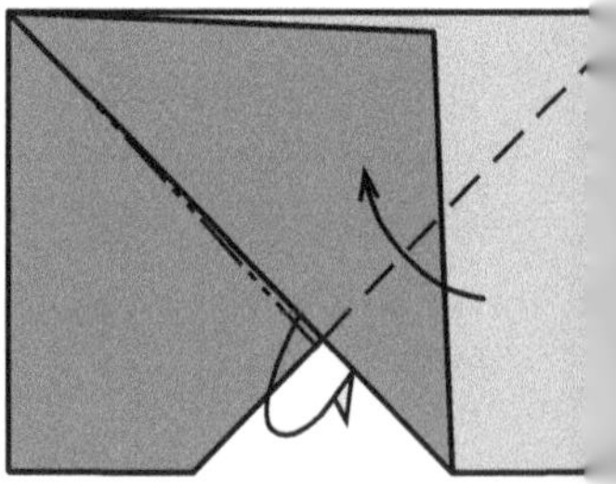

3 Fold the right corner u Turn over and repeat behind.

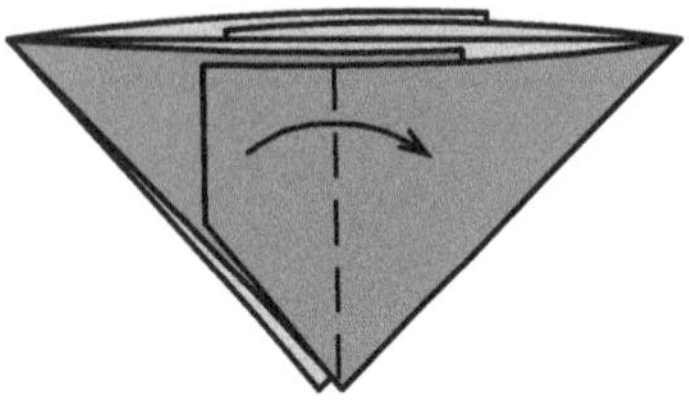

4 Fold the front flap right. Turn over and repeat behind.

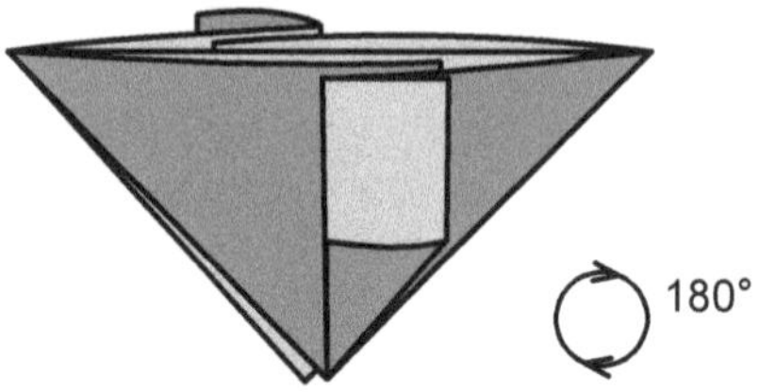

5 Lift these flaps so that they are perpendicular to the right-angled isosceles triangle.

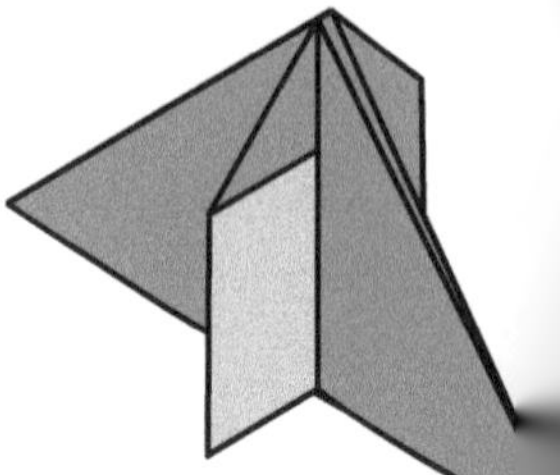

6 Like this. Make a total unit using three colour

Assembly ★★

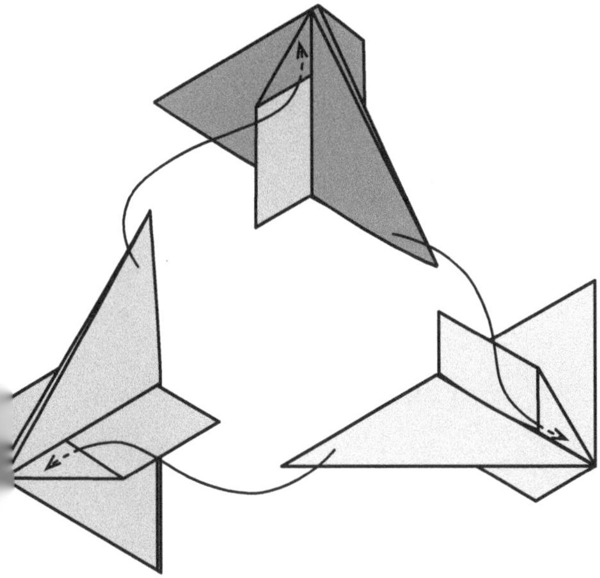

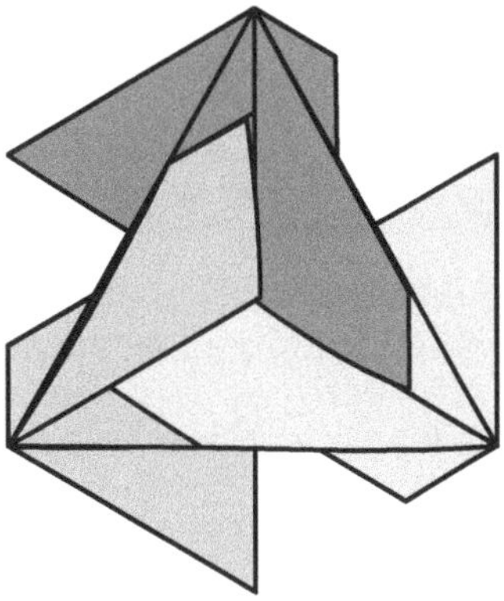

Take three units. Tuck each point into the pocket of the next unit.

2 Continue adding units. Use units of the same colour opposite each other.

3 Gently squeeze the units together to make the assembly secure.

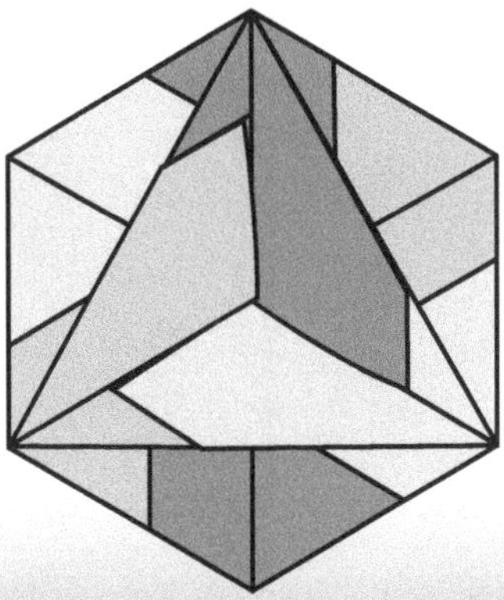

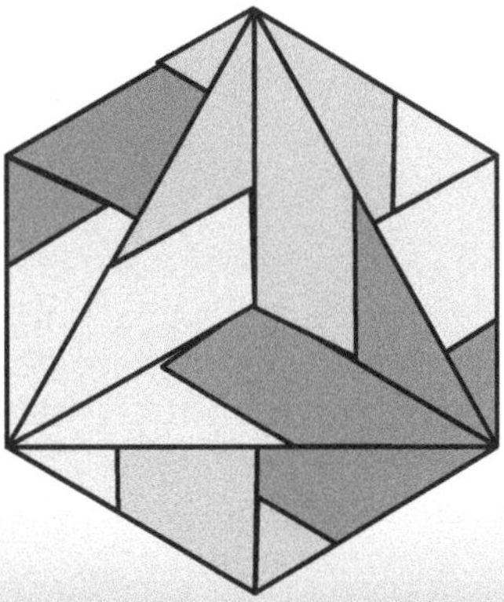

Action

Cup the model in your hands by holding the vertices in your palms. Blow to spin the model.

WXYZ

★★★☆☆
Use 12 squares
Finished model ratio 1.0

WXYZ was the result of experimenting with skeletal cuboctahedra, which can be seen as four intersecting regular hexagons: what happens if the hexagons are transformed into equilateral triangles? This spawned other planar units like my *Blintz Icosidodecahedron* and creations by others in *Planar Modules* [Petty 13] and in Meenakshi Mukerji's website.

Use four colours so that each triangle is a different colour. Sizes from 75mm are effective.

Module ★★★

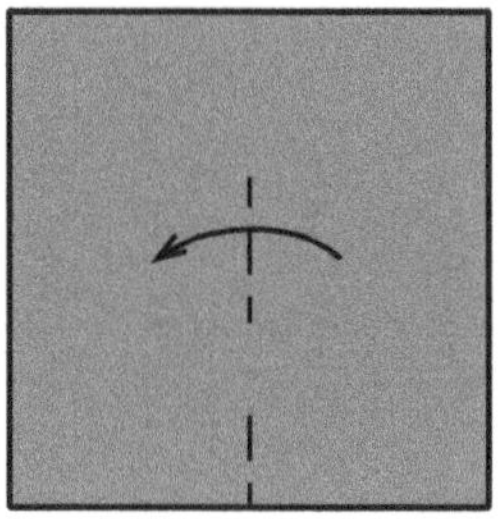

1 Pinch middle and bottom.

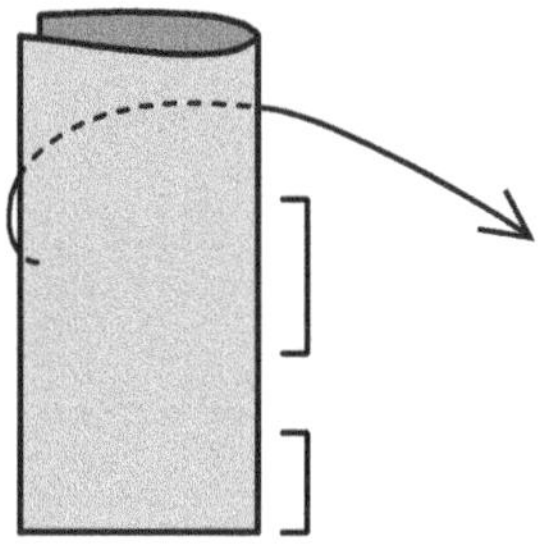

2 Unfold so that the white side is uppermost.

3 Pinch one quarter at bottom right.

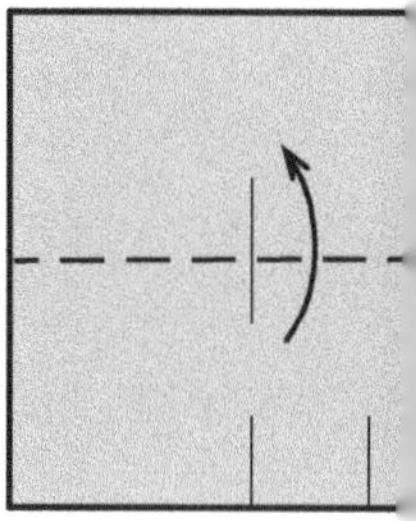

4 Valley in half upwards.

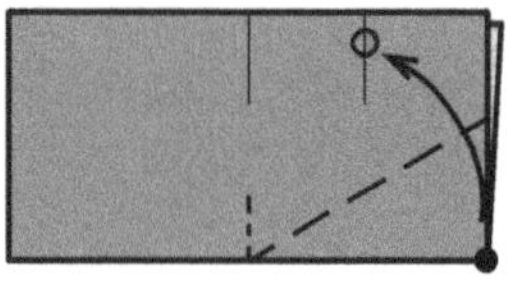

5 Use the mountain at the centre to help locate the fold.

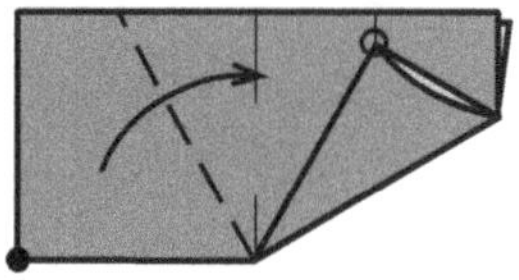

6 Fold the bottom left edge to meet the edge just folded.

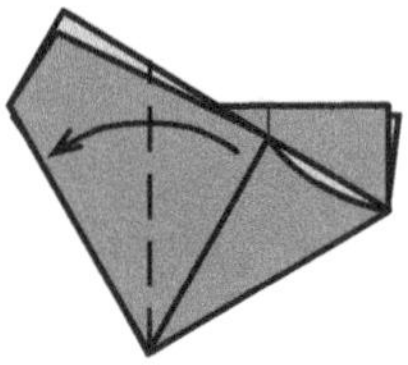

7 Fold the right edge just folded to meet the left edge.

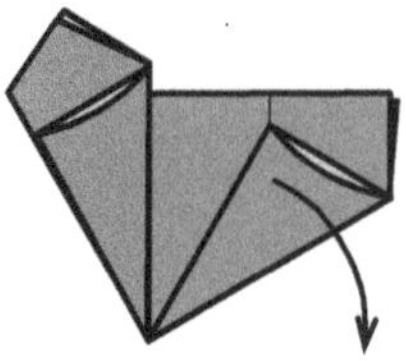

8 Unfold the right flap.

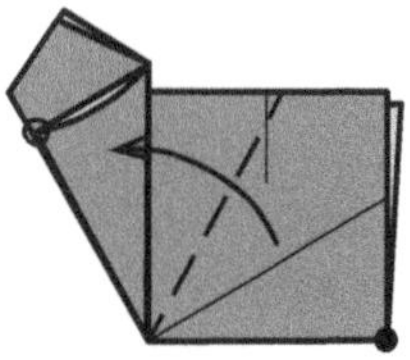

9 Fold the right corner to the left corner.

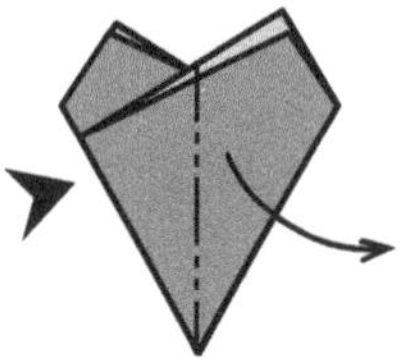

10 Squash the flap to the right.

11 Mountain fold the right flap behind.

12 Fold the left flap the left.

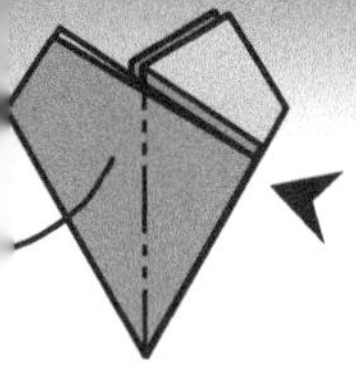

Squash the flap to the right.

14 Squash the flap to the right.

15 Fold down the white flaps against the coloured raw edges.

16 Stand the flaps at right angles.

sembly ★★★

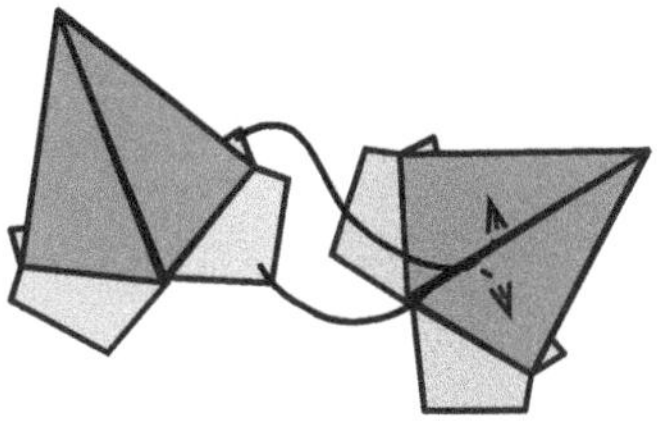

1 Each unit has a pair of pockets on each side. Tuck two flaps of the second unit into one pair of pockets.

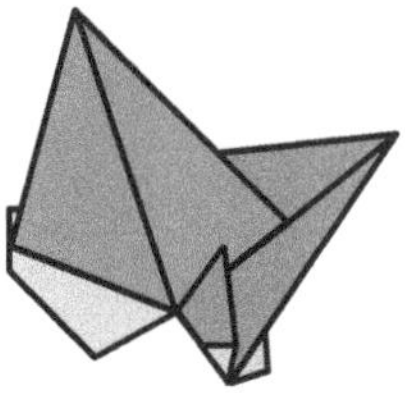

2 Two units joined. Add two more so that you have a ring of four units, each a different colour. Add more units to extend each plane of colour so that you have four triangles, each a different colour.

ction

up the model in your hands with the triangular points in your palms. Blow to spin the odel. You can also put each index finger into opposite pockets and blow.

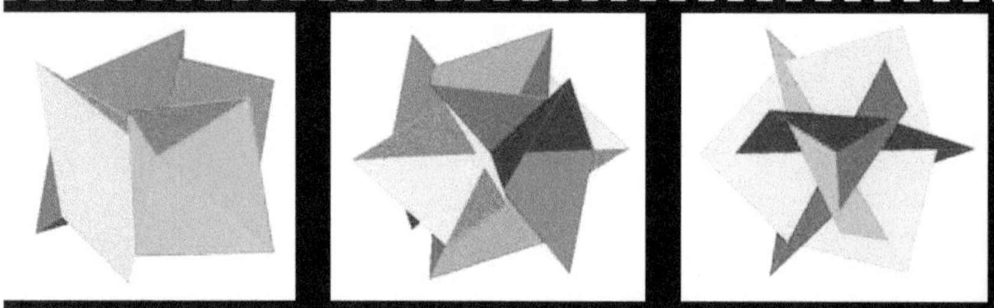

Beak Unit

★★★☆☆

Use 4, 6 or 12 square

Finished model ratio 0.

This single module can make three different planar models: a cube and two kinds of cuboctahedra. It's related to the unit for the *Blintz Icosidodecahedron* [Petty 13]. As always, make sharp and accurate creases. In particular make sure that the folds in steps 4 to 8 go through the centre. You can make more 60° folds in step 1 for more accuracy. You can make some other folds to improve accuracy, but these need some extra creases that show in the fi model. If you find a folding sequence that better suits your paper and folding style, then use it instead of the diagrammed sequence.

Module ★★★

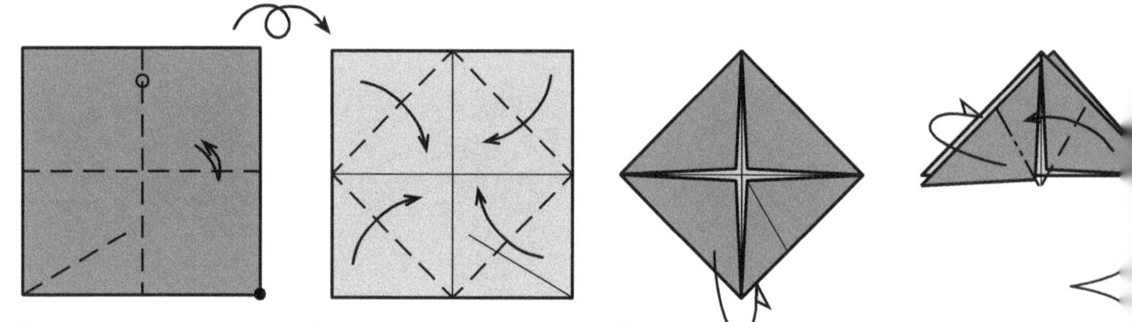

1 Divide a square into a 2x2 grid. Make a 60° fold at the bottom left corner. Turn over.

2 Fold the corners to the centre.

3 Fold the bottom half behind.

4 Use the 60° fol made in step 1 divide the angle thirds at the mi of the base.

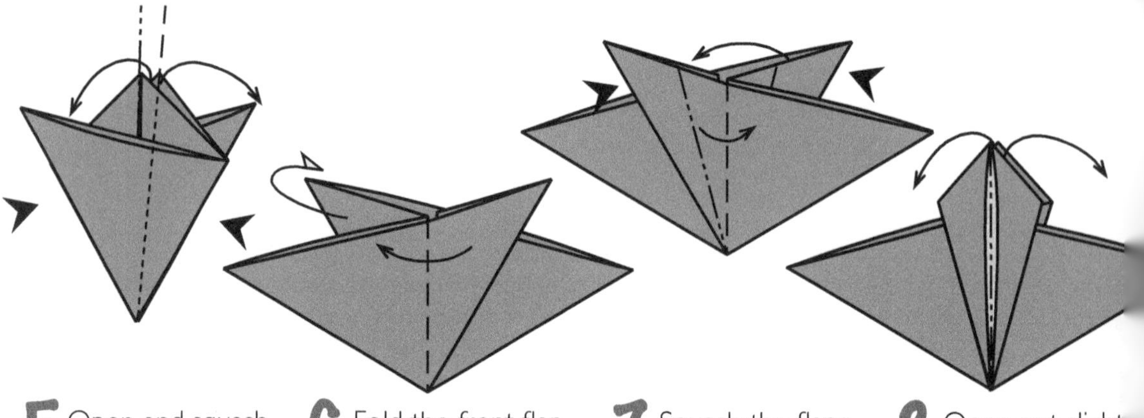

5 Open and squash.

6 Fold the front flap to the left. Repeat behind.

7 Squash the flaps.

8 Open out slight

-piece Cubic and 6-piece Cuboctahedral Assembly ★★★

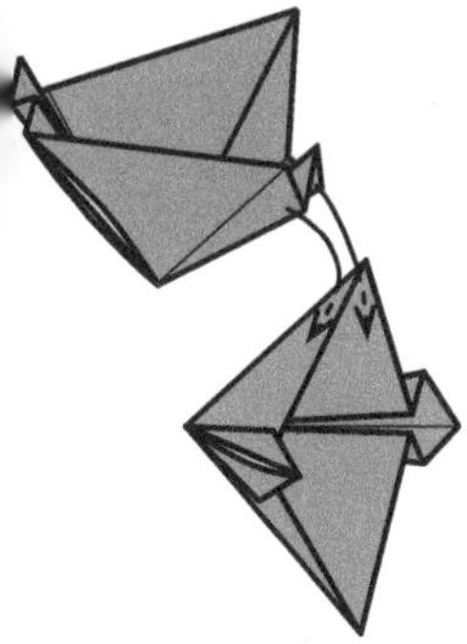

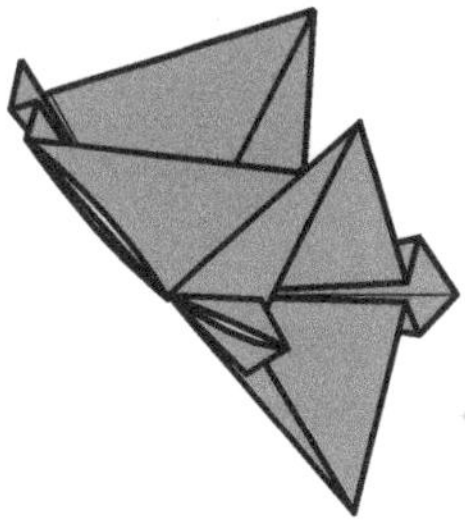

Tuck the tips of the first unit into the pockets of the second unit.

2 Two units joined. To join six units, add more units with mirror symmetry. Three colours work well.

3 To join four units, tuck the tips of the second unit into the pockets of the first unit. Then add the third and fourth units with rotational symmetry. Two or four colours work well.

-piece Cuboctahedral Assembly ★★★

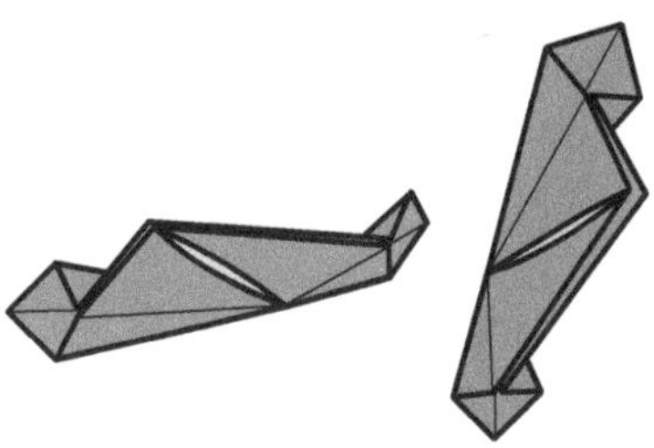

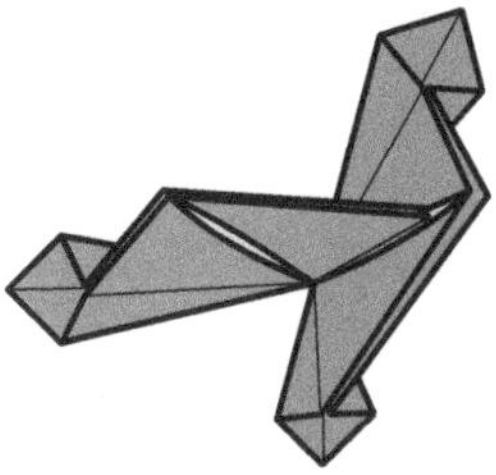

In module step 8, open the module and squash. Tuck the tips of the first unit into the pockets of the second unit.

2 Two units joined. To join twelve units, add more units with rotational symmetry. Four colours work well.

3 Observe the pattern of triangles and squares.

Action

Gently cup opposite vertices in your palms and blow. Or gently hold the the model between the fingertips of each hand and blow. Each method uses an axis of rotation.

Octagon Waterwheel

★★☆☆☆

Use 8 2:1 rectangles
Finished model ratio 0.8

This waterwheel has many variations. This one starts from a conveniently-sized rectangle: one half of a square which means all you need are four squares, ideally two or four colours, each square cut in half.

I have chosen easy fractions in steps 2 and 5: one quarter and one half, respectively. These fractions make the width of the wheel neither too wide nor too thin in step 6.

If you start with rectangles of a different proportion, what fractions will give pleasing results?

Module ★

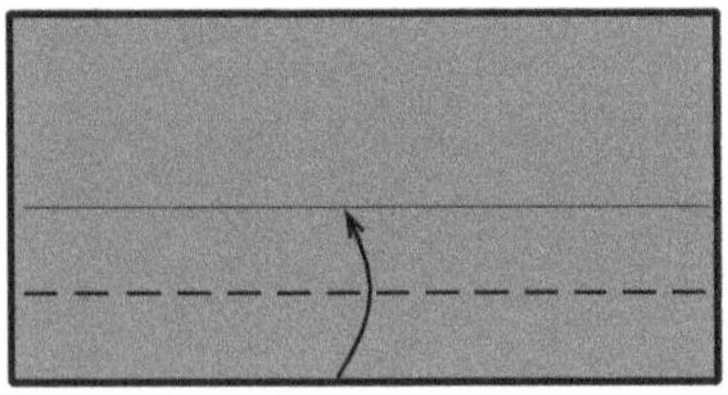

1 Crease in half then fold one quarter up.

2 Turn over.

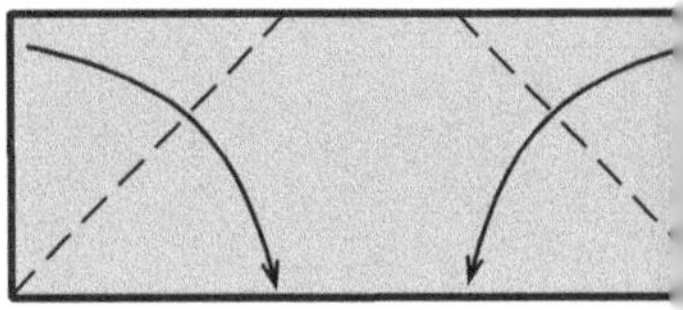

3 Bisect the corners, foldir both layers of paper.

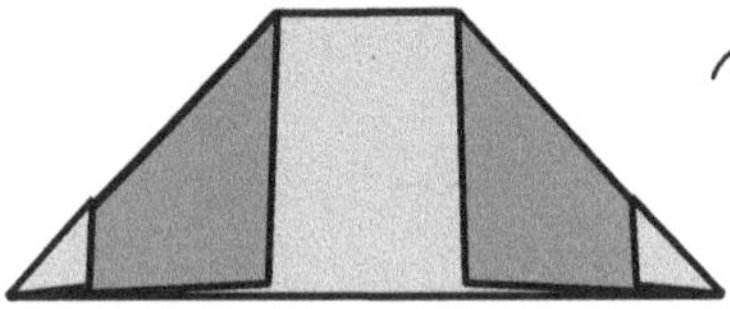

4 Turn over.

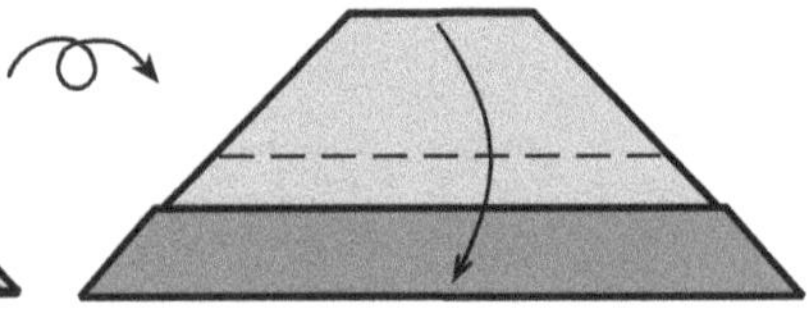

5 Fold in half.

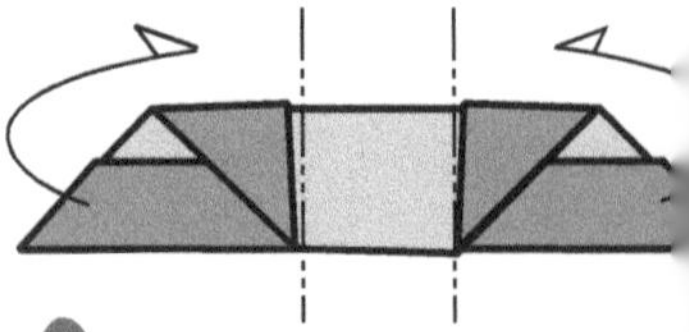

6 Fold flaps at right angles

7 One unit completed. You need eight units; two or four colours work well.

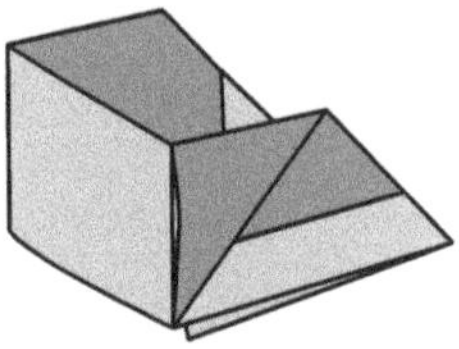

ssembly ★★★

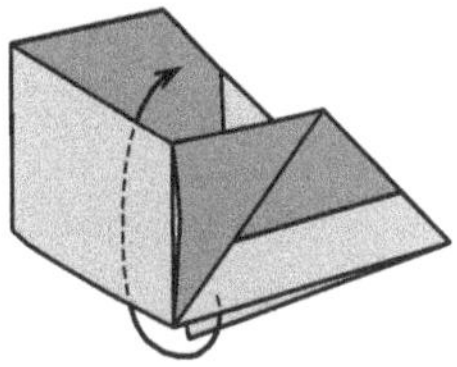

1 Lift up the hidden raw corner flap and hold in place. (The next steps shows this as a convex cone.)

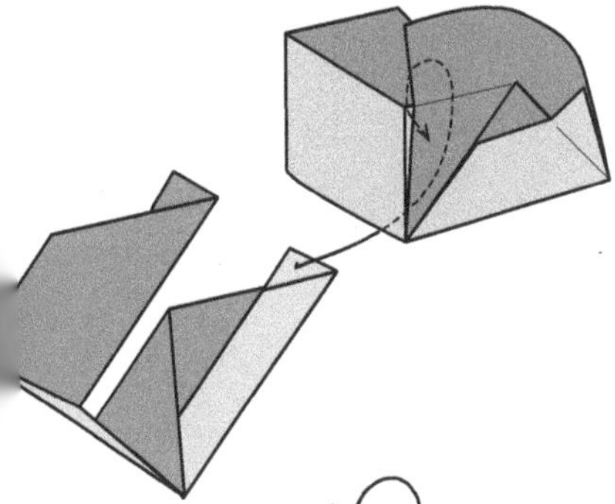

2 Take a second unit and slide its visible raw corner flap through and under the first unit's lifted flap.

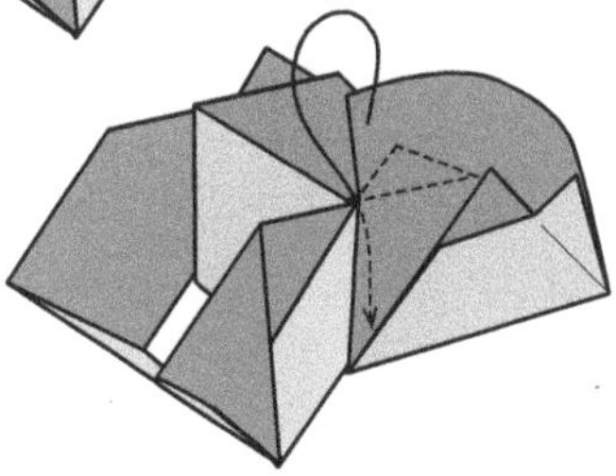

3 Lock the second unit by returning the first unit's lifted flap back to its original position.

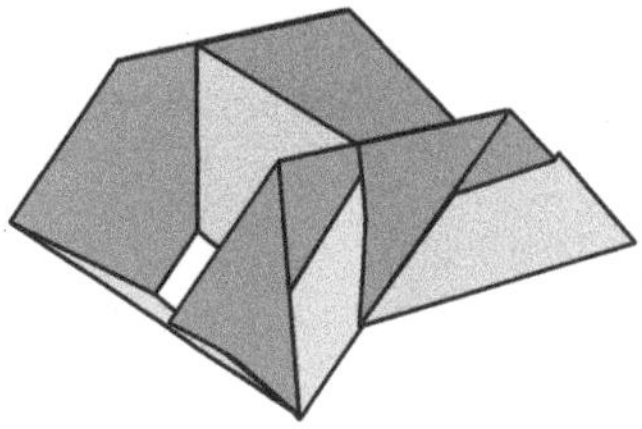

4 Repeat on the other side to join the two units. Continue adding units in a clockwise direction.

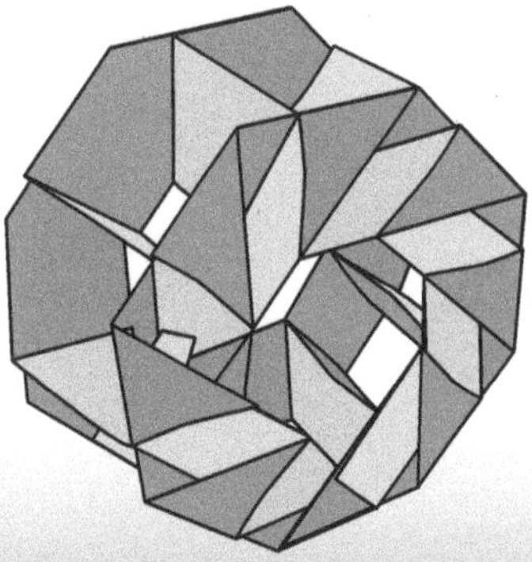

ction

his wheel can roll on a level surface. Furthermore, it will spin very fast when you
t it on a finger (or pencil) and blow.

3D Magic Star

★★☆☆☆

Use 8 silver rectangles e.g. A

Finished model ratio 1.

This 3D slider is inspired by Robert Neale's classic Pinwheel-Ring-Pinwheel. It is also inspire Ray Bolt's *Star Pattern / Frisbee 3D Version* in *British Origami* 62, February 1977; reprinted i *British Origami* 212, February 2002.

This version is the same as the 2D version but introduces a strip of paper in the middle of th unit to make it 3D. Some origami creators call this a river. Adding rivers to 2D units is usually straightforward but choosing convenient fractions and good proportions can be harder.

Module ★★

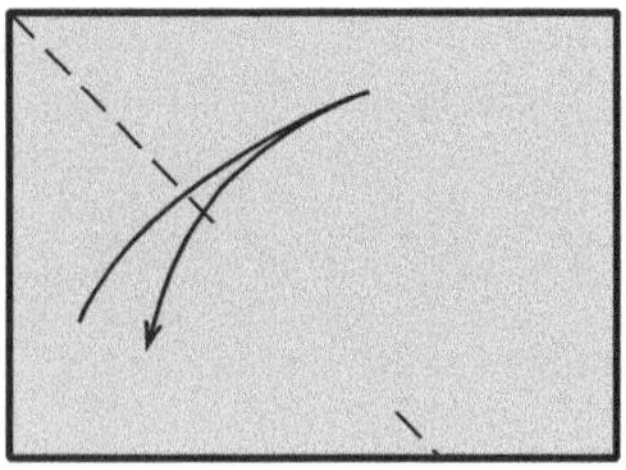

1 Crease the top left corner but leave the middle of the rectangle. uncreased. Make a pinch on the bottom edge.

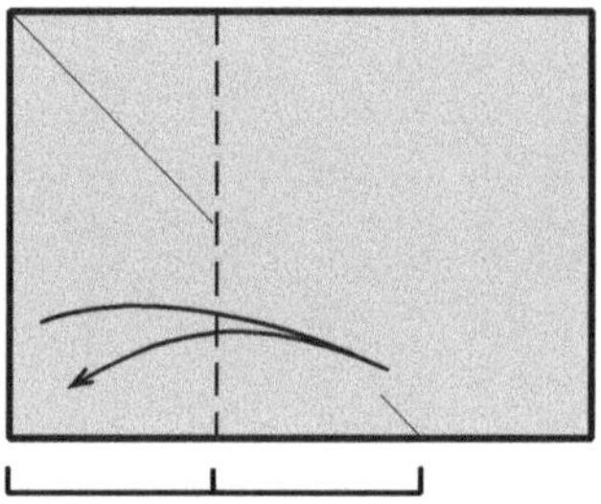

2 This gives the partial diagonal of a square. Fold this square in half and unfold.

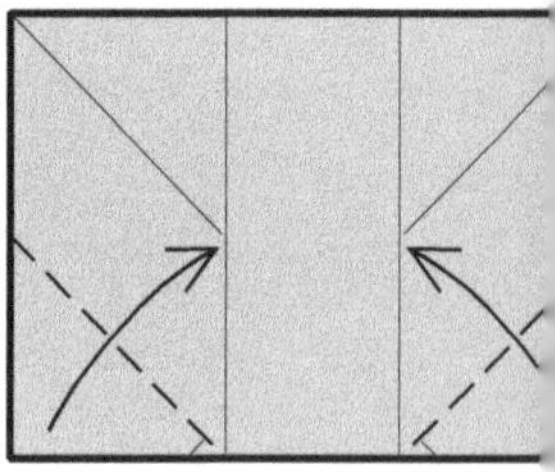

3 Repeat steps 1 and 2 o the right side. Fold the left and right corners u

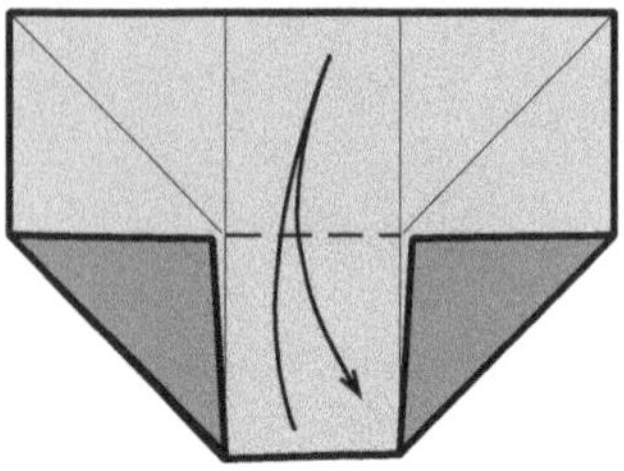

4 Fold in half only between the crease lines.

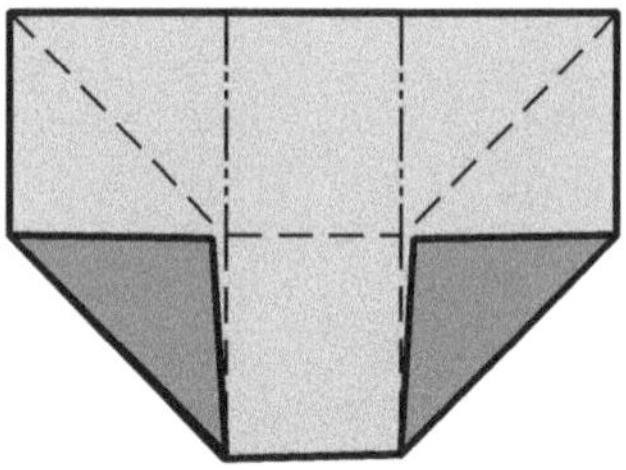

5 Crease and make 3D

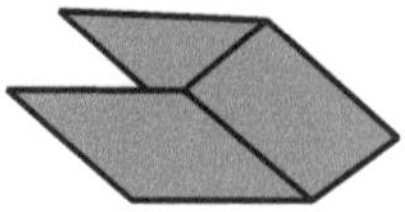

6 Make a total of eight u Two colours work we

Assembly ★★

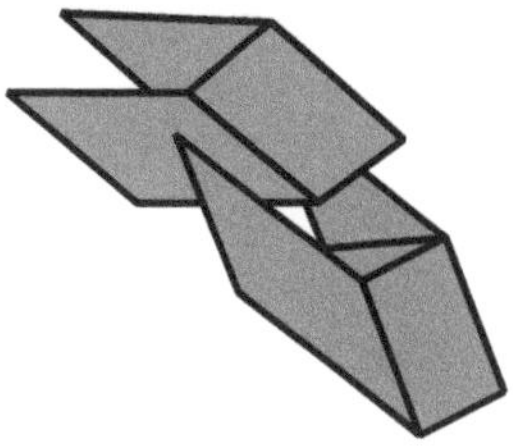

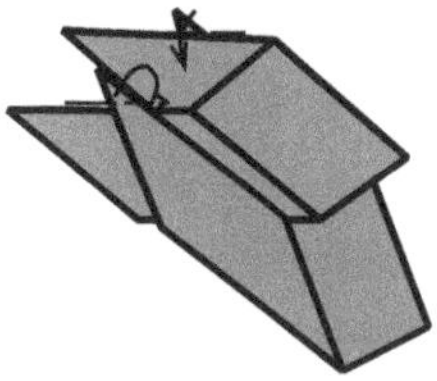

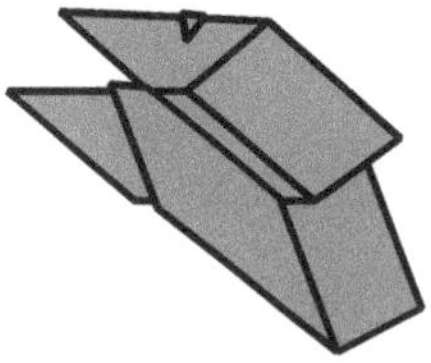

Put the second unit into the seat of the first unit.

2 Tuck flaps in to lock. Test the sliding action and either tighten or loosen the arms, if needed.

3 Continue adding units in an anticlockwise direction. Slide the units to make space if needed.

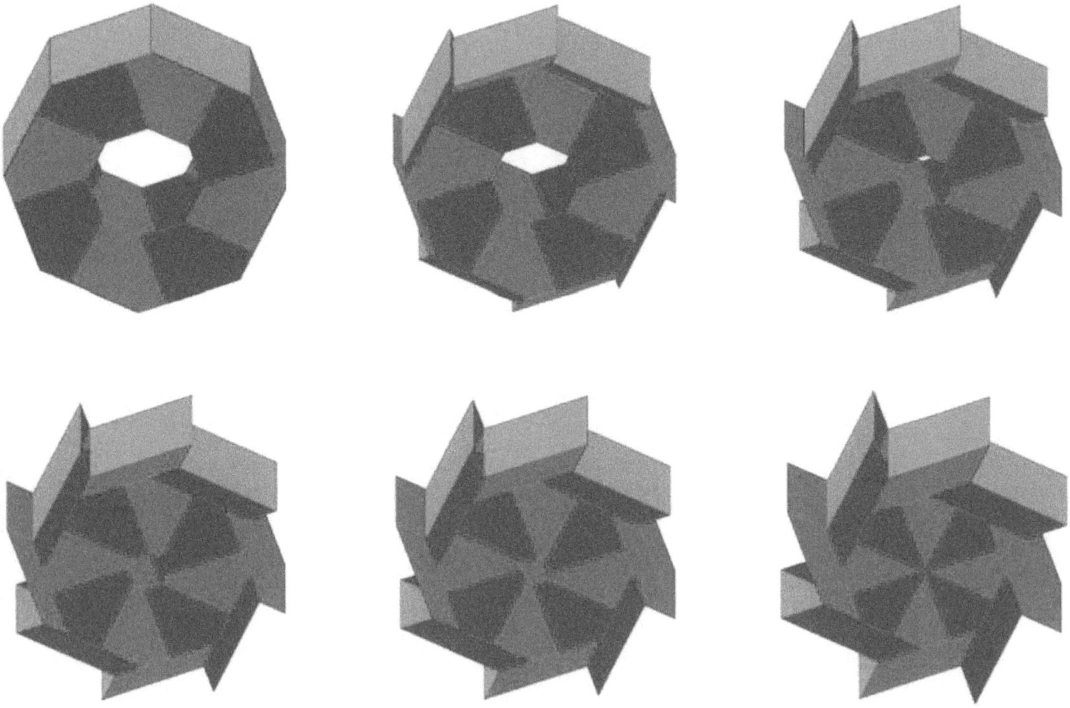

Action

Carefully slide the units to open and close the star.

Octa Slider 3D Wheel

★★★☆☆

Use 8 sheets of A6

Finished model ratio 1.0

You can see that this model is based on the Ribbon Slider. Two or four colours work well.

Module ★★★

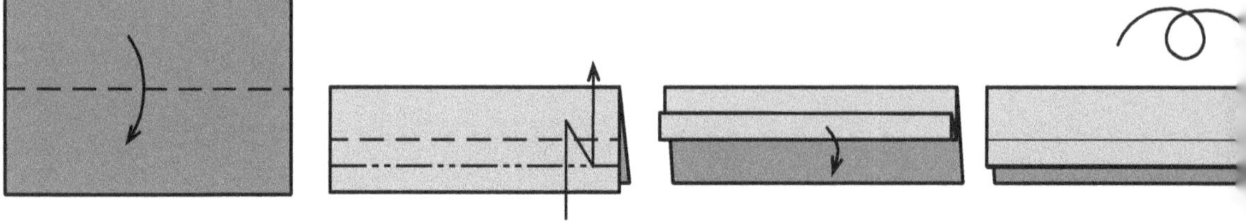

1 Fold in half.

2 Fold the top layer in half and then again.

3 Fold the top flap down.

4 Turn over.

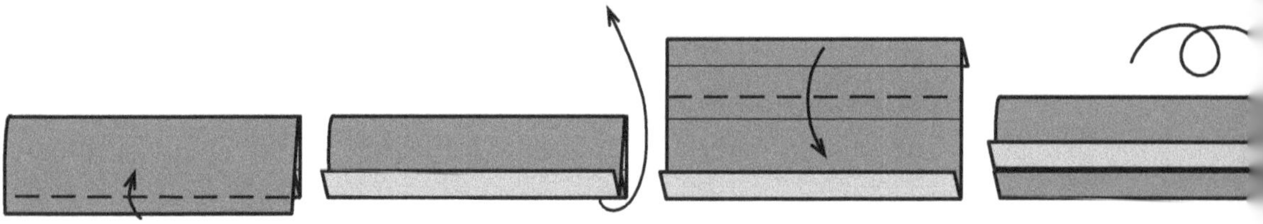

5 Fold the bottom edge to match the back flap.

6 Lift up the back flap.

7 Fold the top folded edge to the raw edge.

8 Turn over. Follo steps 12 to 17 c Ribbon Slider.

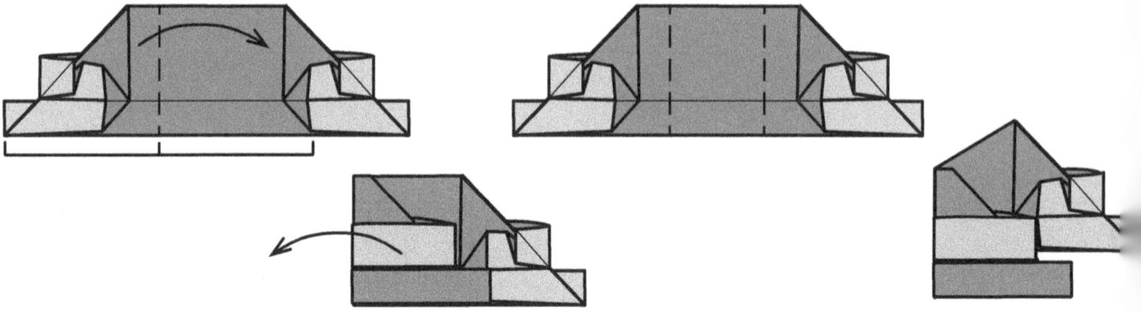

9 Fold in half as shown.

10 Unfold and repeat.

11 Put the flaps at right angles to complete the unit. Assemble like the Ribbon Slide

Chapter 5

3D Shapeshifters

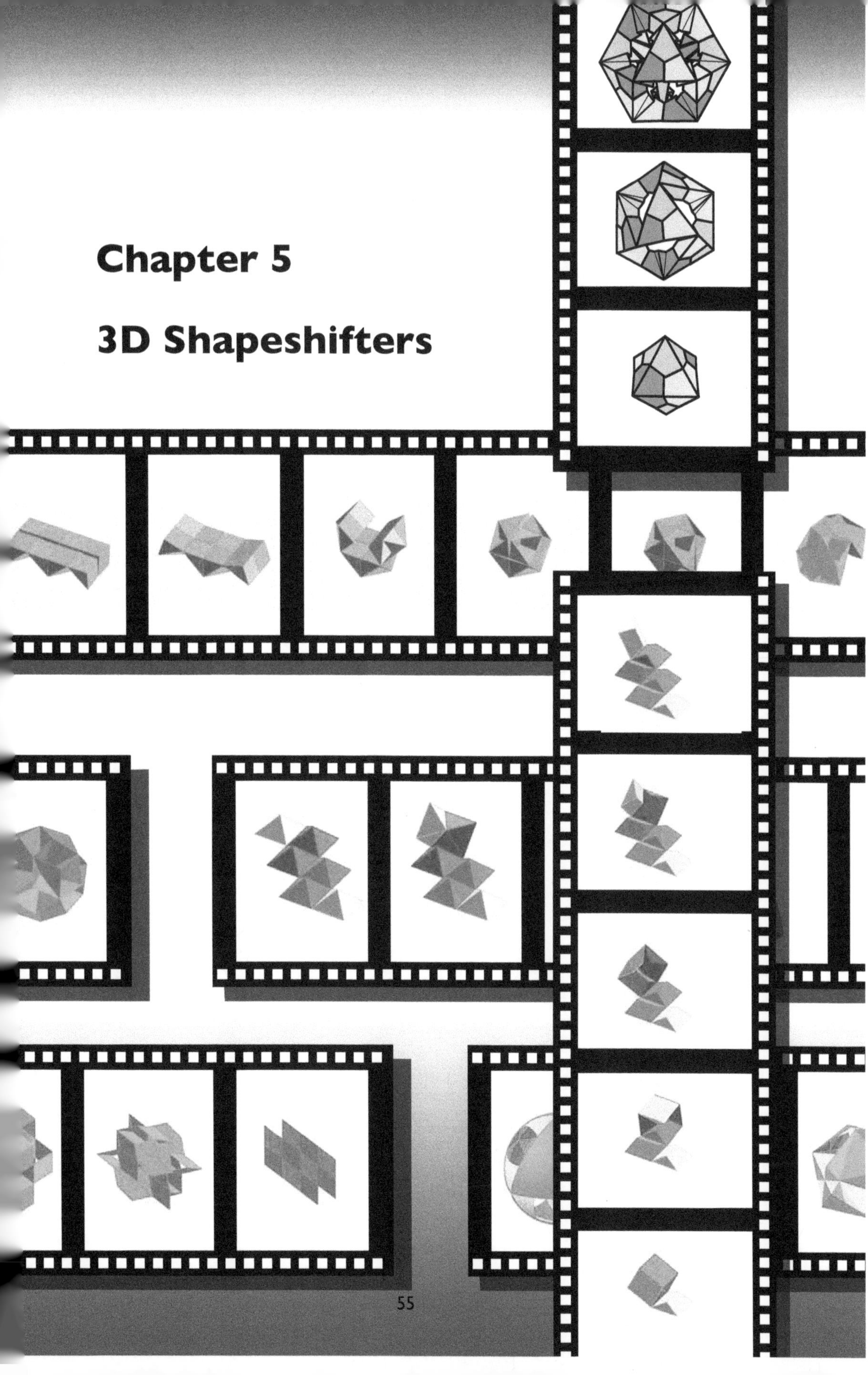

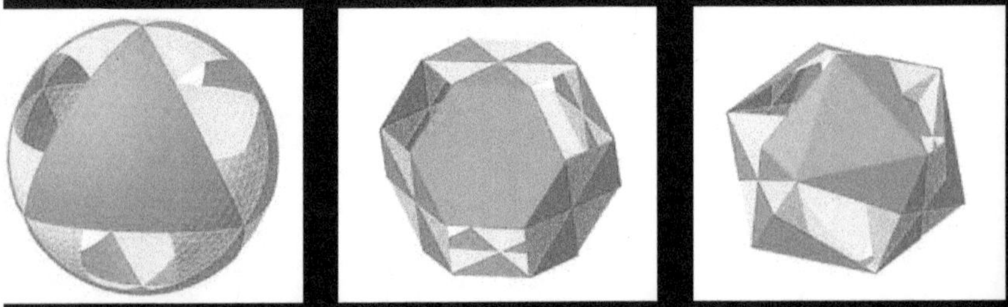

Radioactive Ball

★☆☆☆☆

Use 8 circle

Finished model ratio 1.(

This ball has two actions. It can roll like any other ball but can also be transformed from a truncated octahedron to a regular icosahedron. The name comes from the resemblance t the warning sign for radioactivity. Use circles with a diameter between 10 and 15cm.

You can make module steps 7, 8 or 9 or all steps, depending on which shapes you want to ma Note that the partial icosahedron has left- and right-handed forms.

Module ★

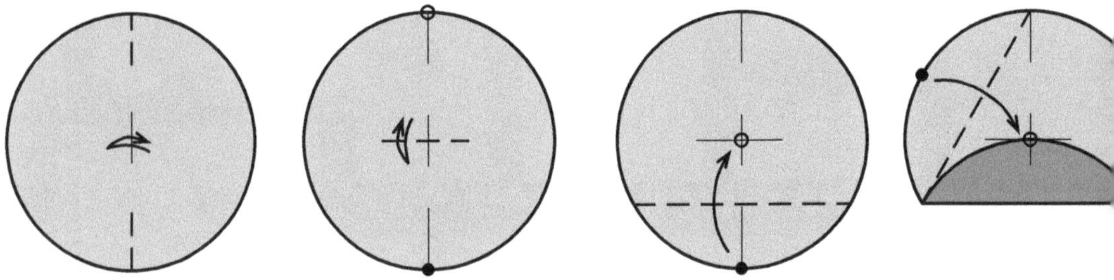

1 Fold in half, only creasing the middle and edges.

2 Fold the ends of the first crease to locate the centre of the circle.

3 Fold to the centre.

4 Fold to the cent The crease runs through the first and third crease

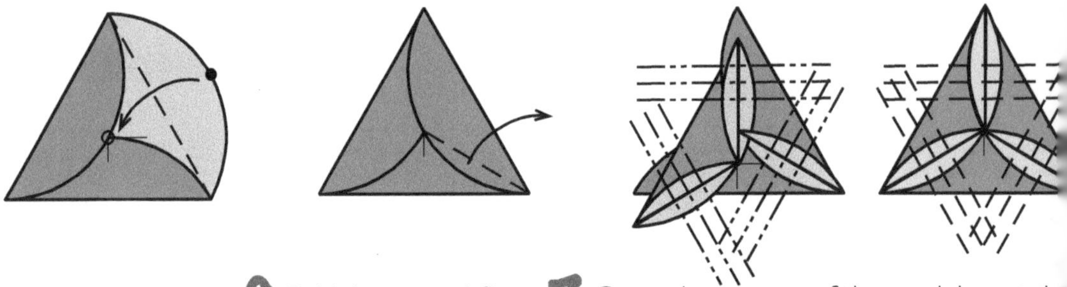

5 Fold to the centre, creating an equilateral triangle.

6 Fold the curved flaps outwards on lines from the centre to the vertices.

7 Curve the corners of the module: put th module near the edge of a table and pu corner down. After repeating several tim the corner will be curved (convex). Mak total of four of the first type of the mod Use a second colour to make four modu of the second type: these are the same a first type but have concave curves.

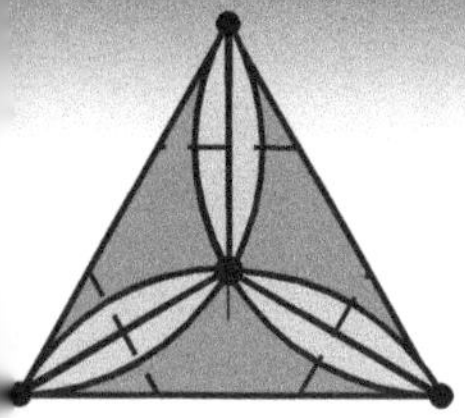

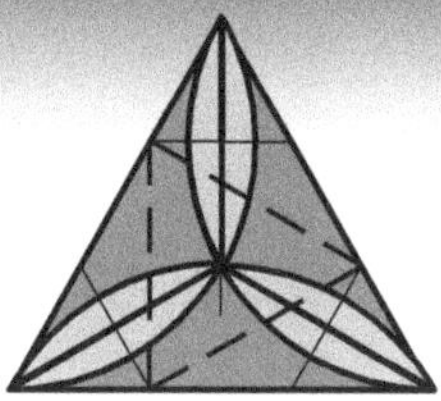

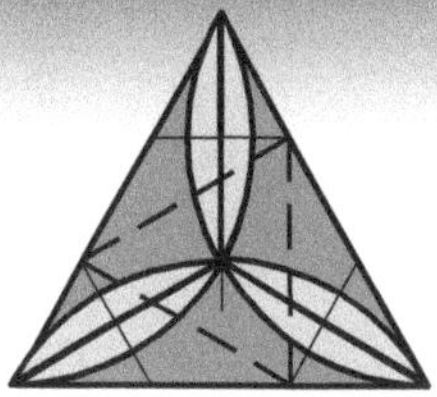

3 Fold the vertices to the centre to make creases for the truncated octahedron.

9 Make folds connecting the alternate vertices of the inner hexagon. Make the same folds on the other first type modules.

10 Make mirror image folds for the second type of unit.

ssembly ★

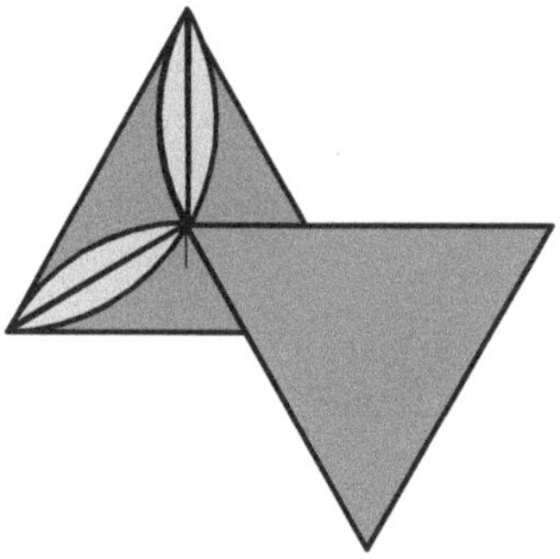

Insert the flaps of the second unit into the first. Turn the second unit over as you do this.

2 Like this. Add two more units.

3 Adding the fifth unit will make the model three-dimensional. Add more to make the ball.

Action

Roll the ball or reinforce some of the creases to make a truncated octahedron, a regular icosahedron or some intermediate shape.

Jitterbug

★★★☆☆

Use 12 squares
Finished model ratio 1.0

Buckminister Fuller gave the name Jitterbug to this transformation. I first came across the Jitterbug in Amy C. Edmondson's *A Fuller Explanation* (1987). As I didn't have dowels and four-way rubber connectors, I made several cuboctahedra that worked as Jitterbugs but were not very reversible. Some were from paper and others from drinking straws and elastic thread. My first unit, Triangle Unit, was partly successful as a Jitterbug. A better result came from using three units per triangular face (or one unit per vertex) which was published in 2000.

This improved version is slightly harder to assemble but has a stronger lock. The sequence is pleasingly rhythmical and all steps have location points. The pleats in step 18 form the spring that make the model return to its cuboctahedral shape.

Module ★★★

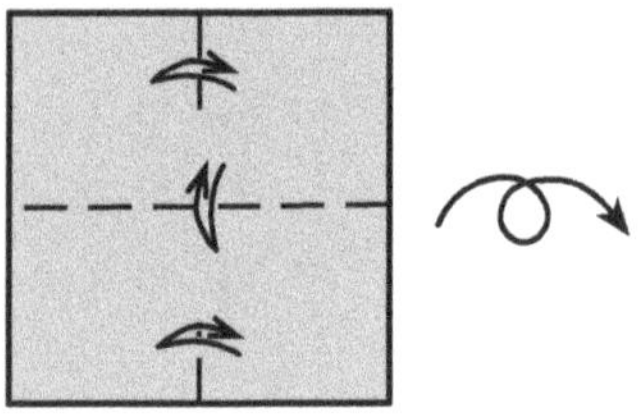

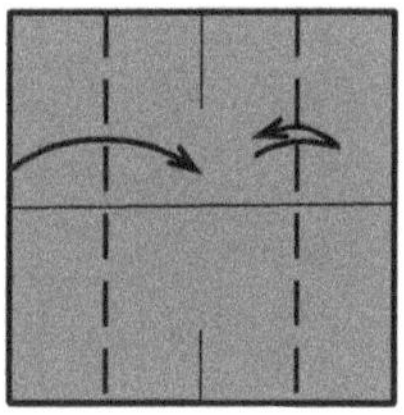

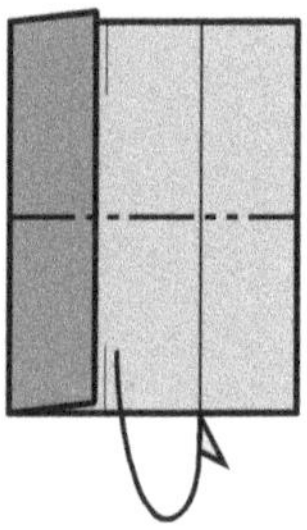

1 Make two folds in half. Do not crease the middle for the vertical fold.

2 Fold the sides to the centre. Unfold the right flap.

3 Fold the bottom half be

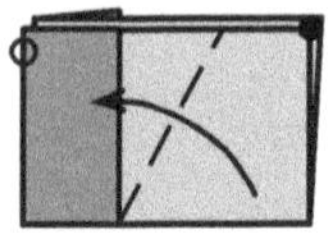

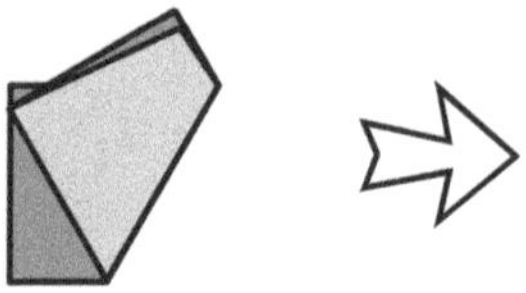

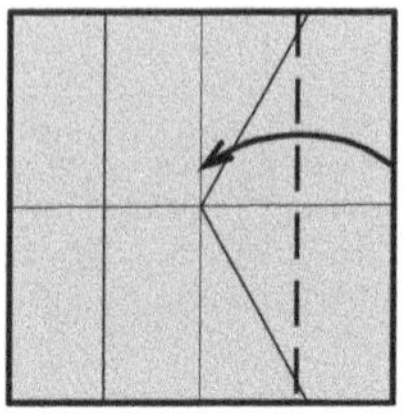

4 Fold the top right corner to lie on the left edge whilst the fold starts from the original centre of the square.

5 This creates a 60 degree angle. Unfold.

6 Fold the right edge to th centre.

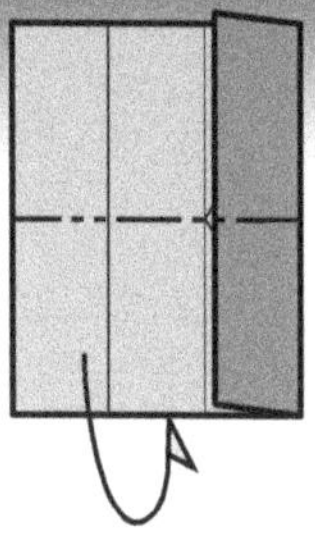

Fold the bottom half behind.

8 Repeat step 4 and unfold, leaving the right flap folded.

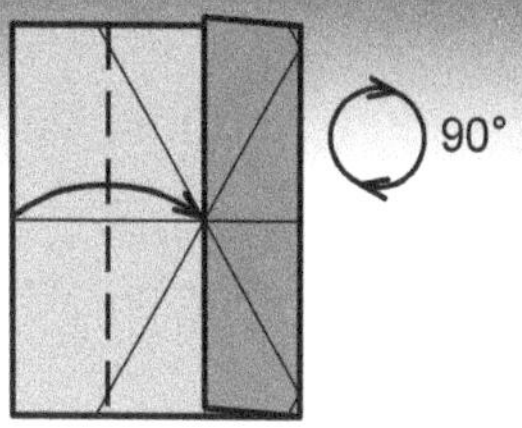

9 Fold the left edge to the centre and then rotate the model 90°.

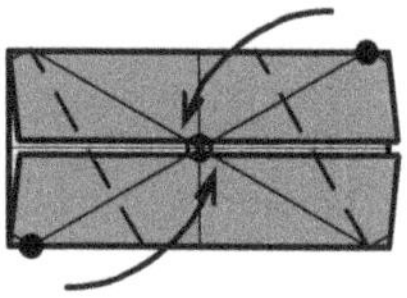

Bisect the 60° angles. Note that the filled circles are the ends of the crease, not the corners.

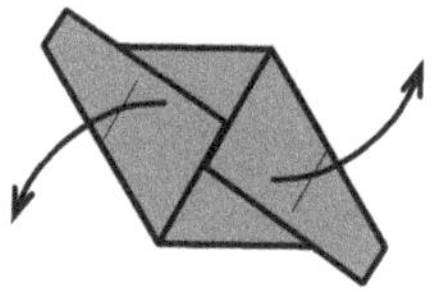

11 Unfold.

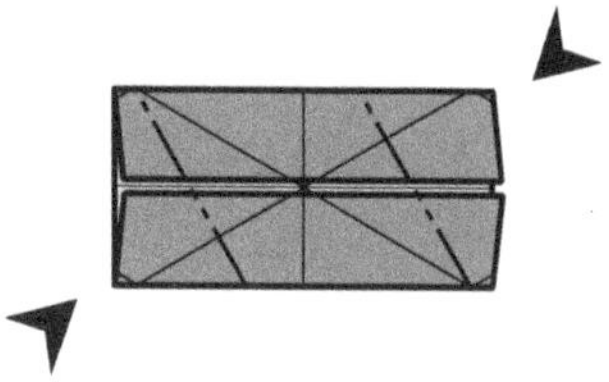

12 Reverse fold. You will need to extend the existing creases slightly.

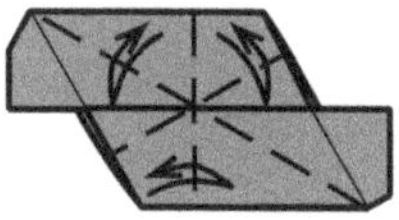

Reinforce three creases.

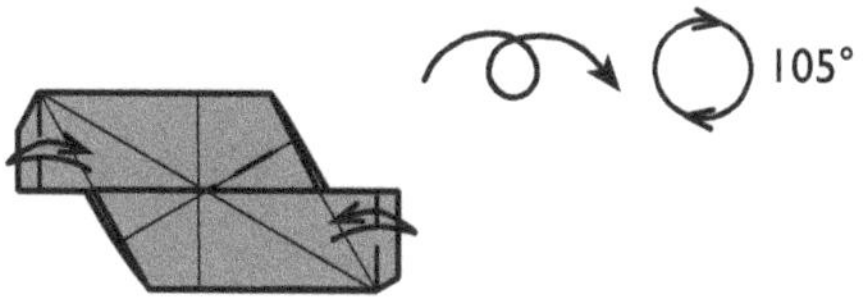

14 Precrease the ends. These creases make the third sides of partial equilateral triangles.

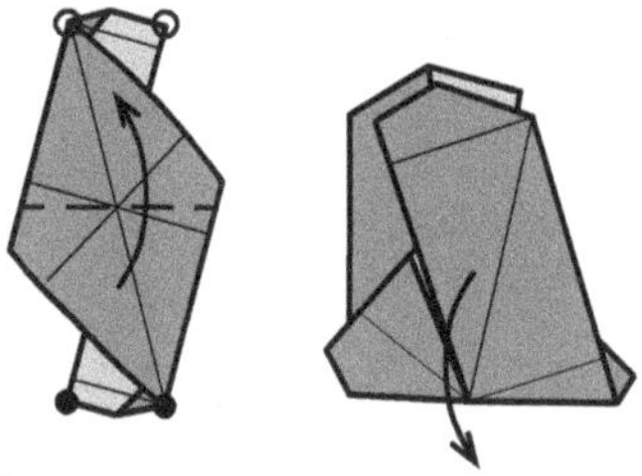

15 Fold in half using the given location points. Unfold.

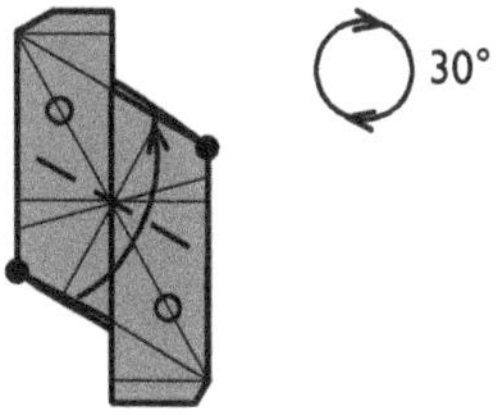

Fold in half using the given location points.

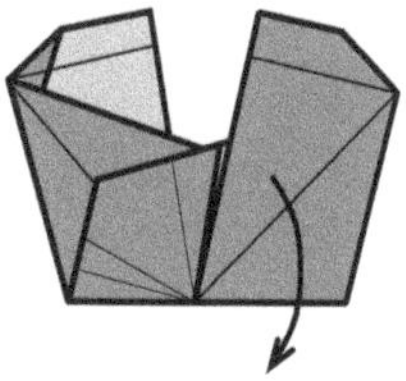

17 Unfold.

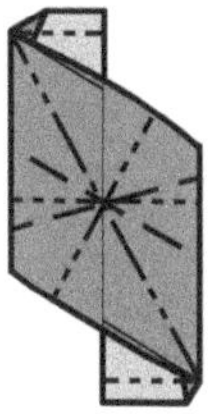

18 Reinforce the creases and make a total of 12 units: three colours works well.

Assembly ★★★

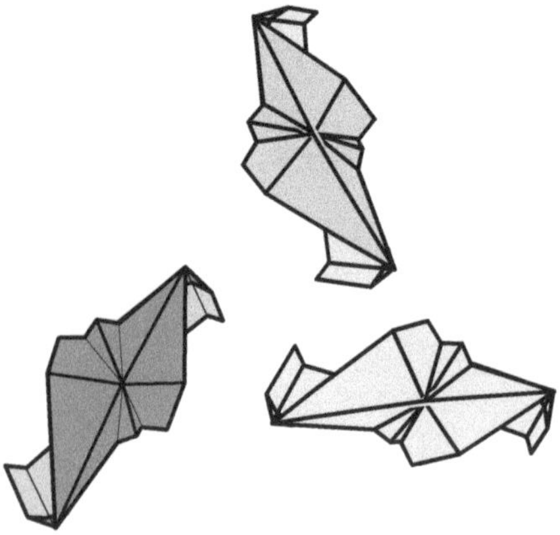

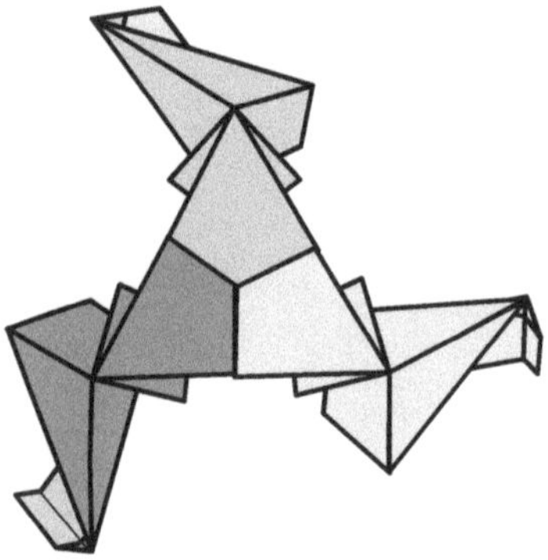

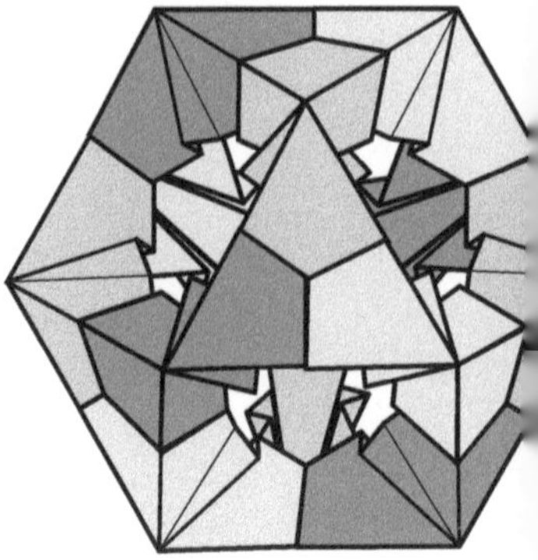

1 Join three units to make a triangle. Use the creases from step 14 to lock the units.

2 Continue adding units to make more triangles. If you are using three colours, note that the order of colours on each face is not always the same.

3 Remember to make square "holes" with four triangles. As you make t cuboctahedron, the vertices have colours th form three rings

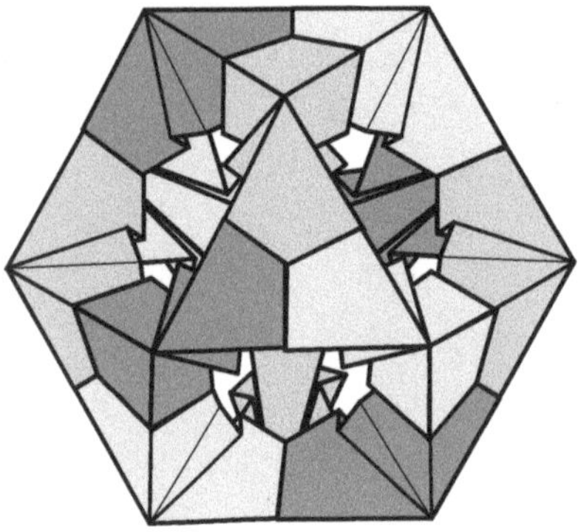

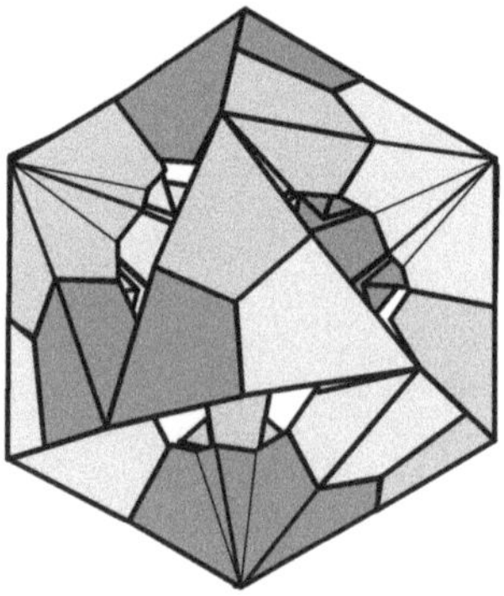

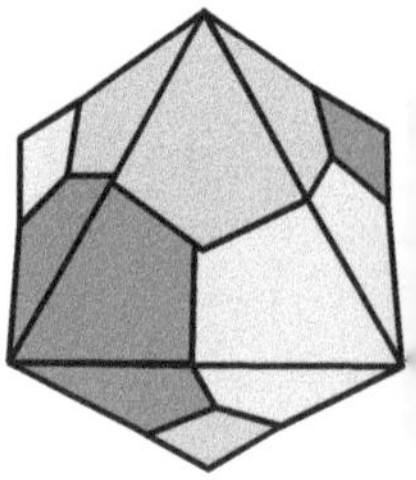

1 Push near and far faces together, faces will twist as they approach…

2 …the circumference will buckle, making an icosahedral configuration …

3 …continue pushing to form an octahedron.

Action

Pressing the opposite faces of Jitterbug transforms it from a truncated cuboctahe to an octahedron — let go and it springs back to its original shape.

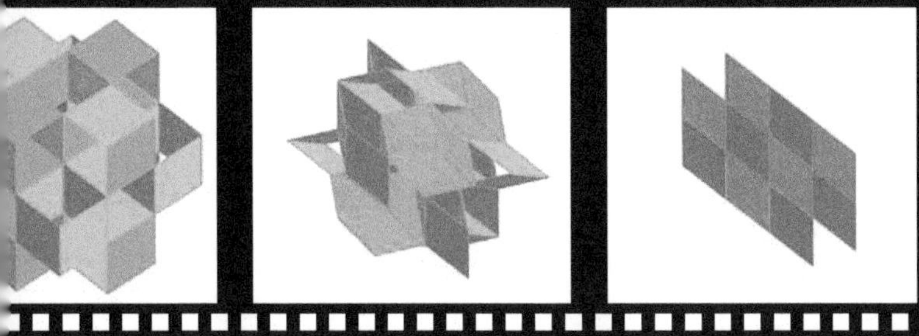

Petrie-Coxeter Honeycomb

★★★★☆

Use 9 squares

Finished model ratio 1.5

first found Petrie-Coxeter Honeycomb described in David Well's Curious and Interesting Geometry. After trying many different way of making the honeycomb using paper, glue and .gnetic modelling kits, it turned out that an existing and well-known module had the right mbination and types of flaps and pockets for the honeycomb: the Sonobé unit with reverse ds developed by Mitsonobu Sonobé, Toshie Takahama, Tomoko Fusé, Kunihiko Kasahara and hers. This versatile unit makes familiar modular origami polyhedra like cubes, triakis octahedra d triakis icosahedra. Use 75mm origami paper or memo cube paper up to about 10cm.

ssembly ★★★

is table summarises the possible modules defined their orientation (left- or right-handed), flap ection (mountain or valley) and colour (1 or 2). e numbers specify the numbers of each module eded for a given assembly.

ere is how to start the 3:1;1 assembly. Follow the ble or follow the pattern of the honeycomb: ernate the units and holes, and alternate the lours of the units.

	Colour 1		Colour 2	
	L	R	L	R
M	LM1	RM1	LM2	RM2
	1			
V	LV1	RV1	LV2	RV2
			1	

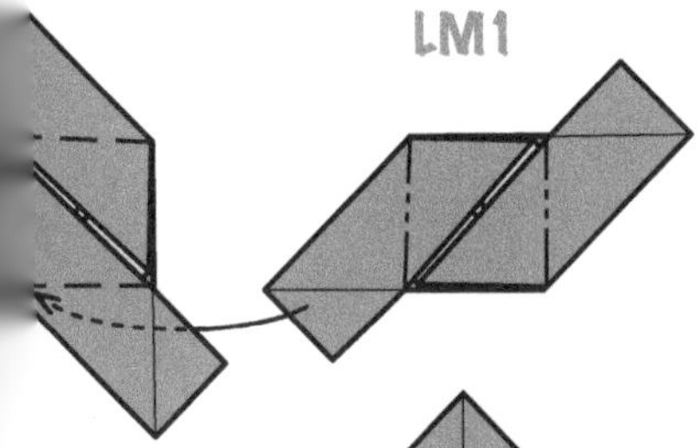

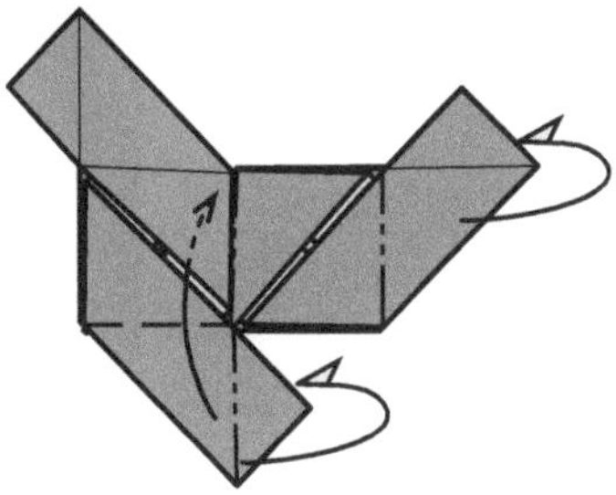

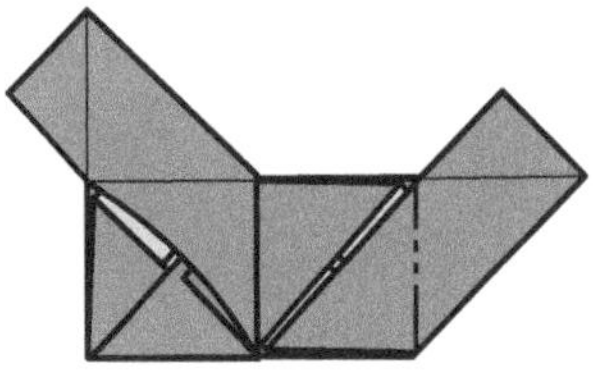

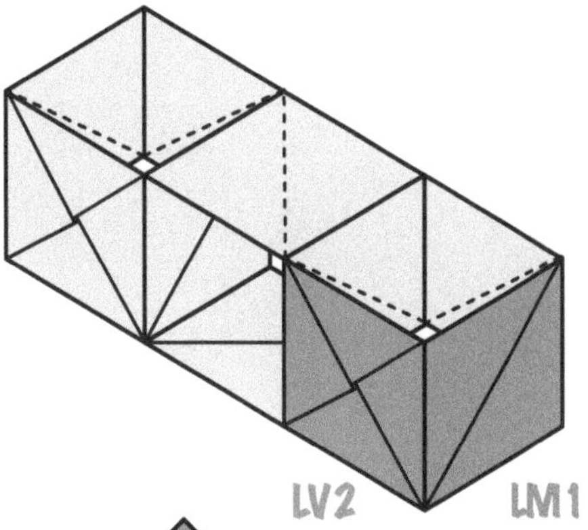

Insert the flap of the first unit into the second unit.

2 Fold the tip of the second unit behind. Insert the whole flap into the pocket of the same unit.

3 Mountain fold the first unit to be at right angles to the second unit.

4 You've made the first part of the 3:1:1 assembly. All other assemblies use these types of locks.

Left-handed Mountain and Valley Modules ★★★

Colour 1

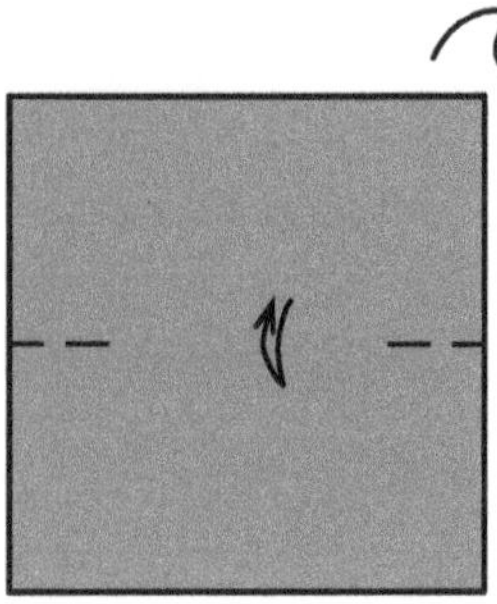

1 Pinch midpoints. Turn over.

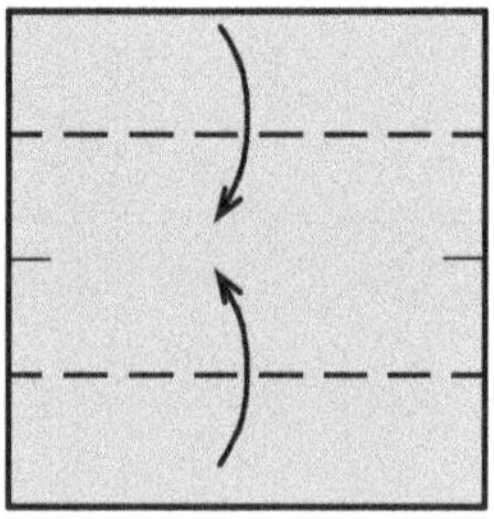

2 Fold top and bottom edges to the middle.

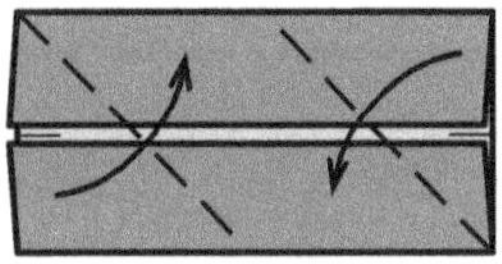

3 Fold the bottom left corner and the top right corner to the middle.

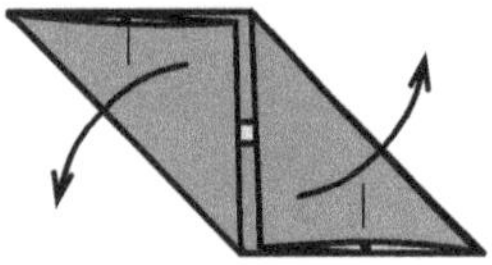

4 Unfold.

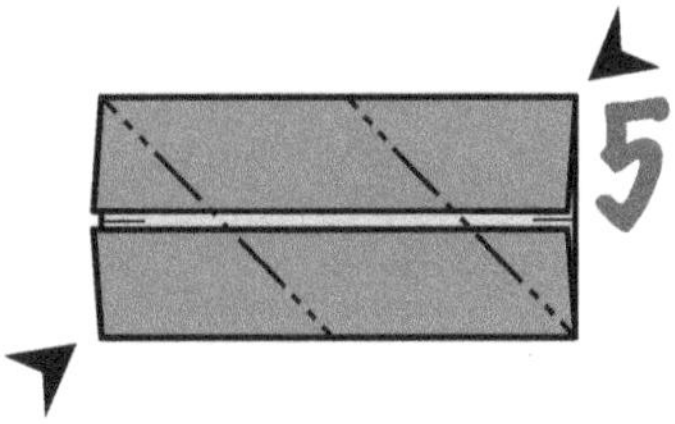

5 Inside reverse fold using the existing creases.

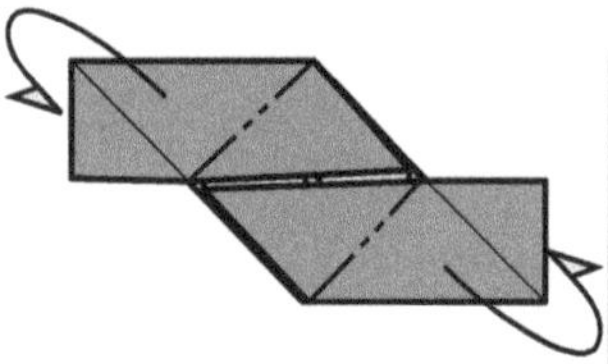

LM1

6 Mountain fold.

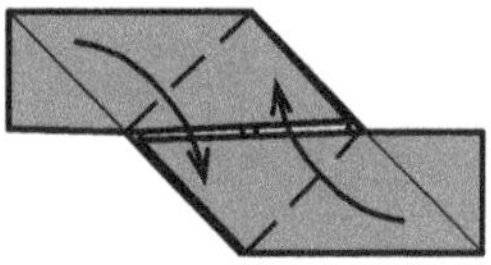

LV1

6 Valley fold.

Colour 2

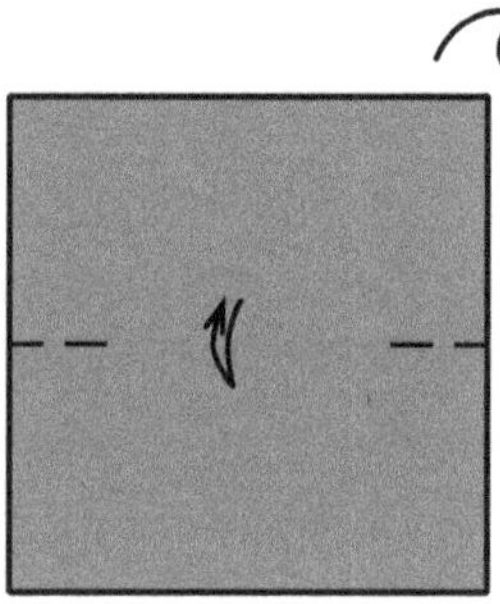

1 Pinch midpoints. Turn over.

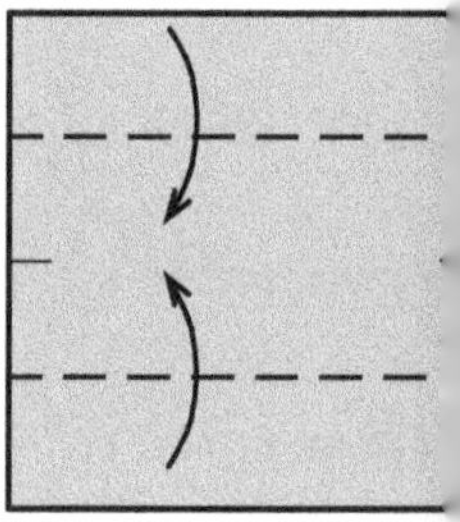

2 Fold top and bott
edges to the midd

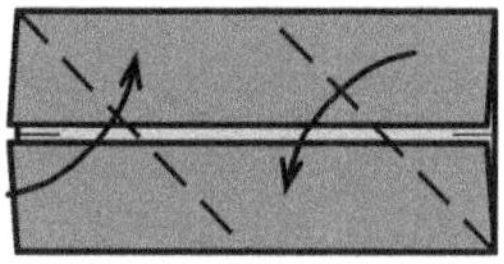

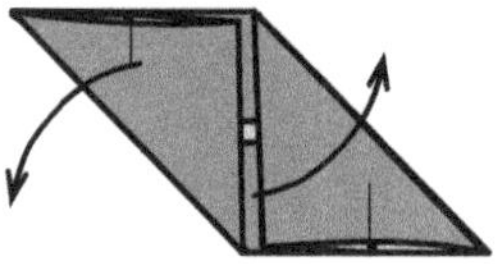

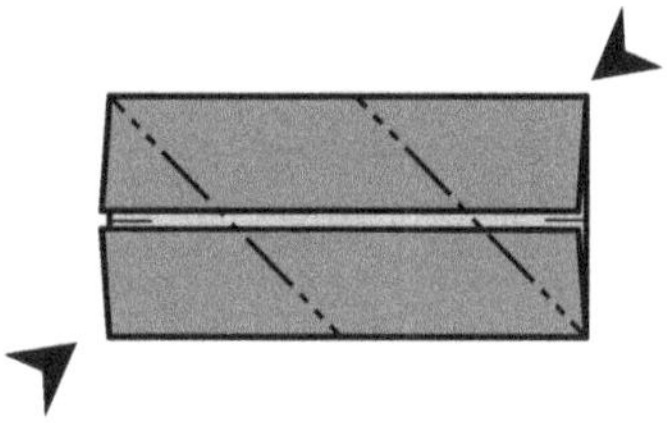

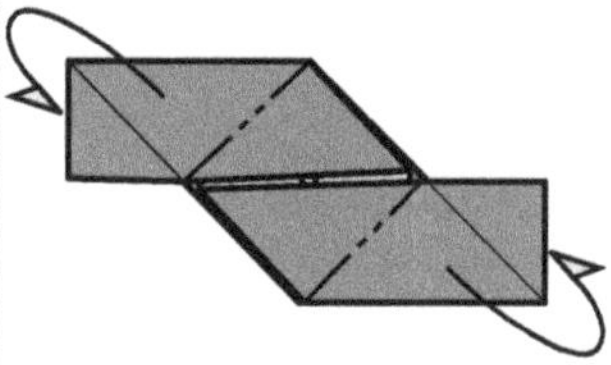

LM2

6 Mountain fold.

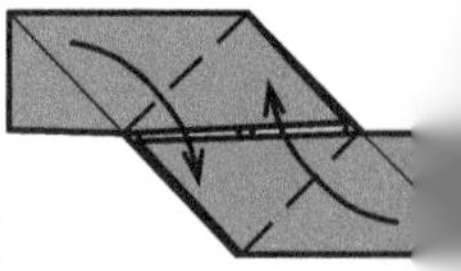

LV2

6 Valley fold.

ght-handed Mountain and Valley Modules ★★★

lour 1

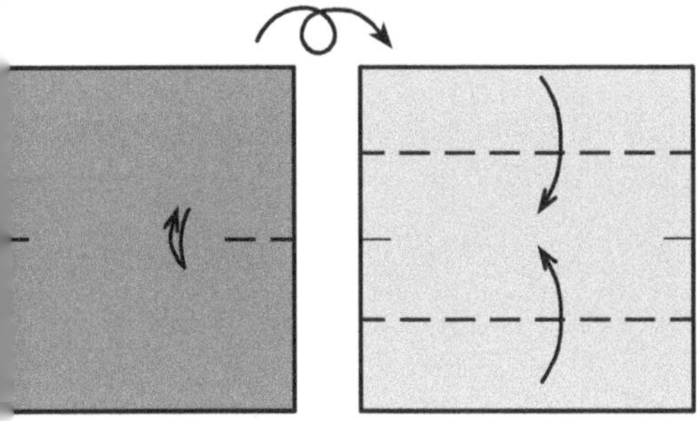

Pinch midpoints. Turn over.

2 Fold top and bottom edges to the middle.

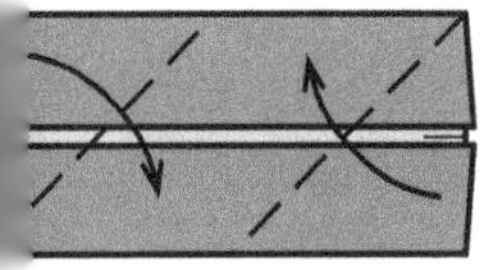

3 Fold the bottom right corner and the top left corner to the middle.

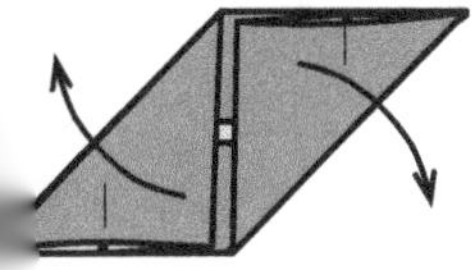

4 Unfold.

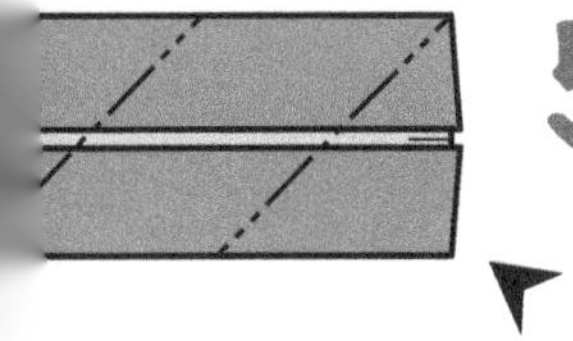

5 Inside reverse fold using the existing creases.

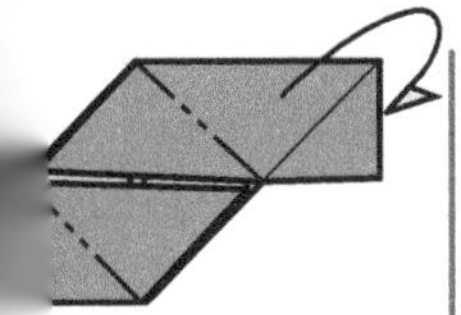

RM1

ountain fold.

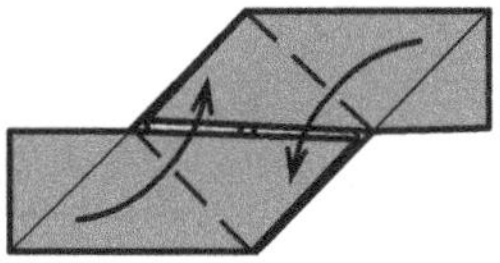

RV1

6 Valley fold.

Colour 2

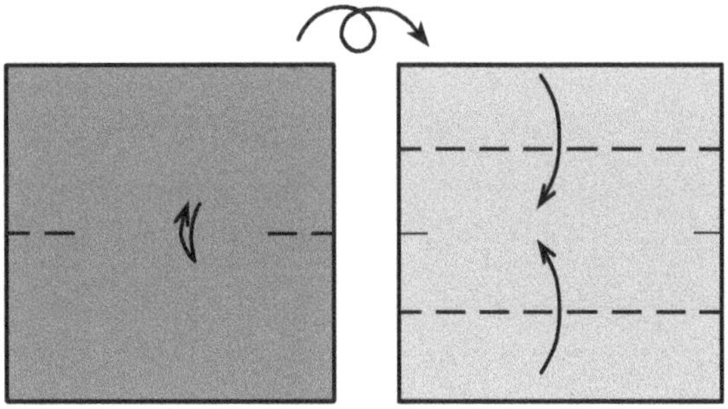

1 Pinch midpoints. Turn over.

2 Fold top and bottom edges to the middle.

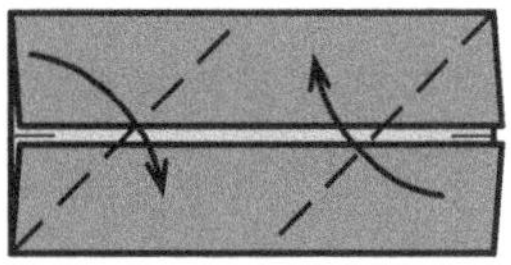

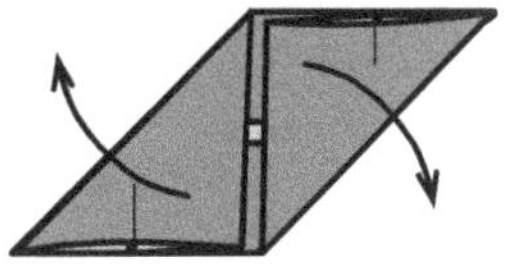

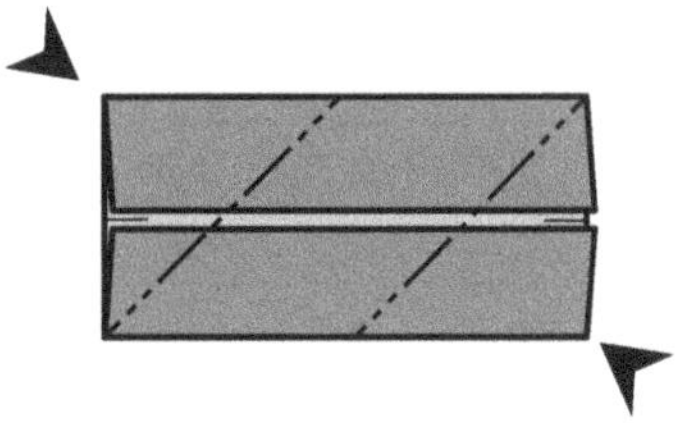

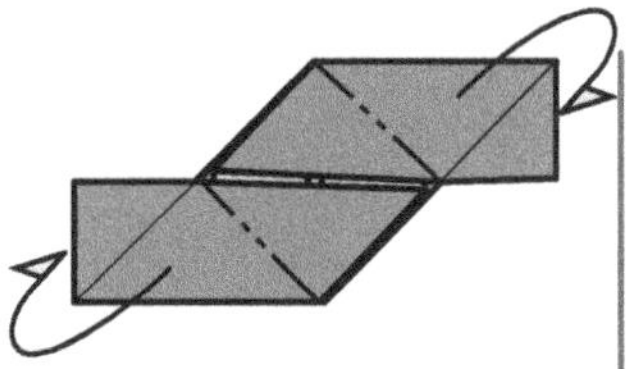

RM2

6 Mountain fold.

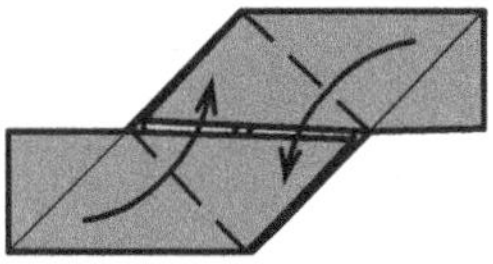

RV2

6 Valley fold.

3:1:1 Assembly ★★

	Colour 1		Colour 2	
	L	R	L	R
M	LM1	RM1	LM2	RM2
	4	2		
V	LV1	RV1	LV2	RV2
		2	4	

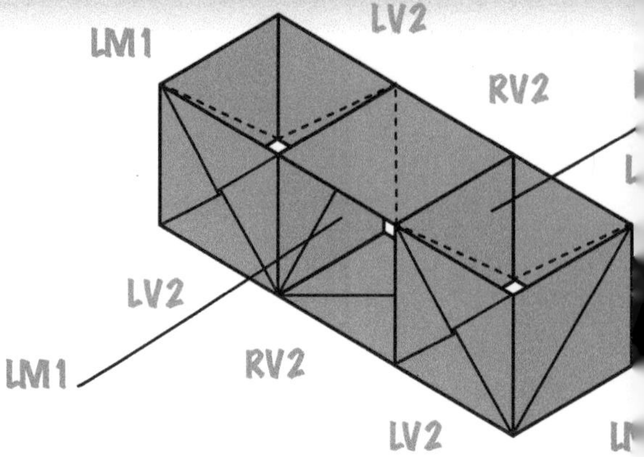

3:3:1 Assembly ★★★

	Colour 1		Colour 2	
	L	R	L	R
M	LM1	RM1	LM2	RM2
V	LV1	RV1	LV2	RV2
		4		

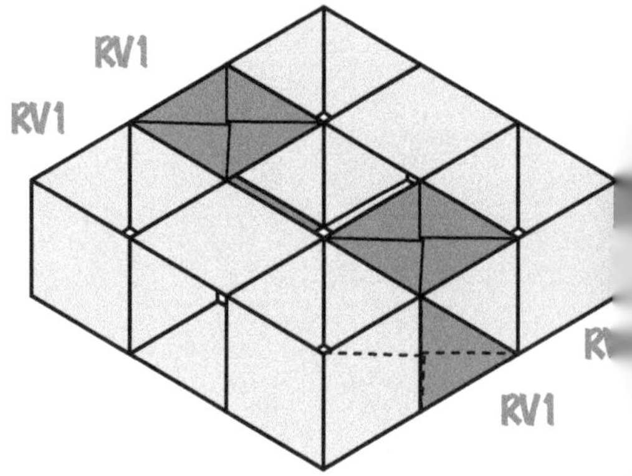

Connect two 3:1:1 assemblies with four RV1 modules. You will need to undo some of the flaps in the LV2 modules to connect the RV1 modules.

3:3:3 Assembly ★★★★

	Colour 1		Colour 2	
	L	R	L	R
M	LM1	RM1	LM2	RM2
	4			
V	LV1	RV1	LV2	RV2

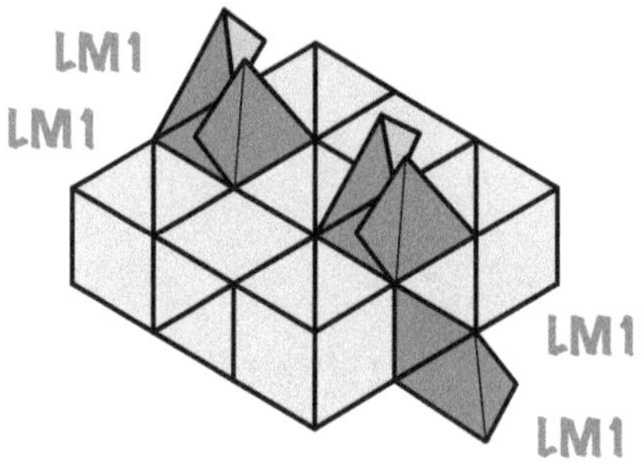

Connect two 3:1:1 assemblies with four LM1 modules to make the middle layer. Make and connect modules in the top and bottom layers.

3:3:3 Top Layer ★★★★

	Colour 1		Colour 2	
	L	R	L	R
M	LM1 4	RM1	LM2	RM2
V	LV1	RV1 2	LV2 6	RV2

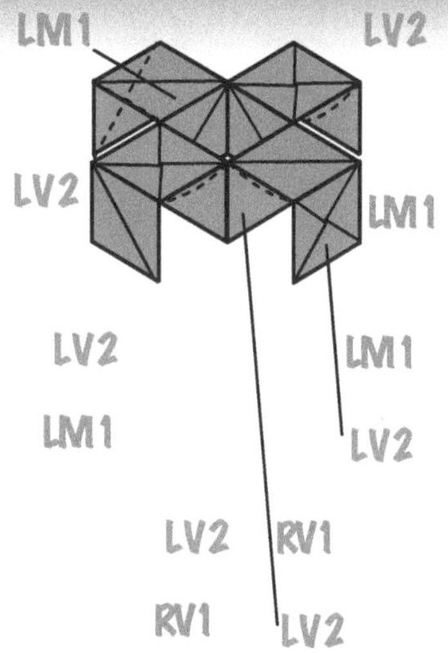

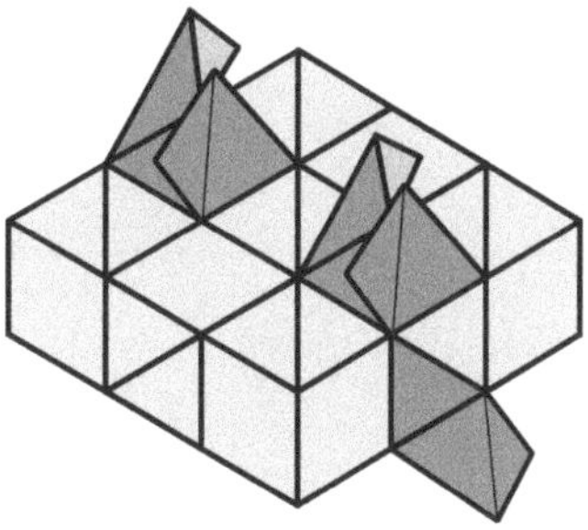

3:3:3 Bottom Layer ★★★★

	Colour 1		Colour 2	
	L	R	L	R
M	LM1 4	RM1	LM2	RM2
V	LV1	RV1 2	LV2 6	RV2

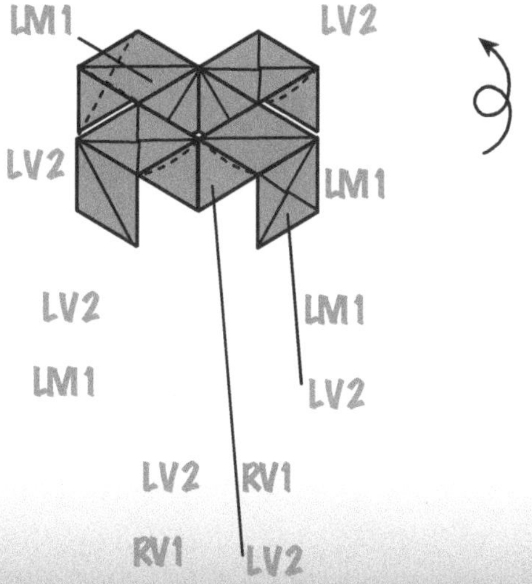

Action

Press on the opposite edges of cube and the square holes will flex into different kinds of rhombuses. You can do to this each set of edges, separately or together. In fact the whole assembly can be flattened. Do not keep it flattened or it will lose its elasticity.

Collapsoid

★★★☆☆

Use 30 sheets of A6

Finished model ratio 1.0

Like the Petrie-Coxeter Honeycomb, the Collapsoid discovered by Jean Pederson is described in David Wells' Curious and Interesting Geometry. My Collapsoid is a bit thick but does spring back into the starting shape. You may find that you can save some time by making four units from A4 paper: precrease a 4 by 4 grid and fold the diagonals of each A6 pane.

Module ★★★

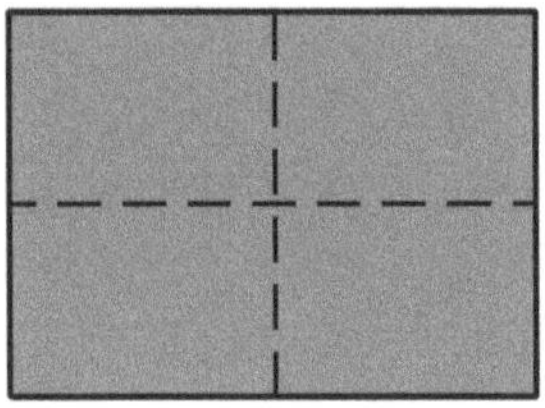

1 Crease a 2 by 2 grid.

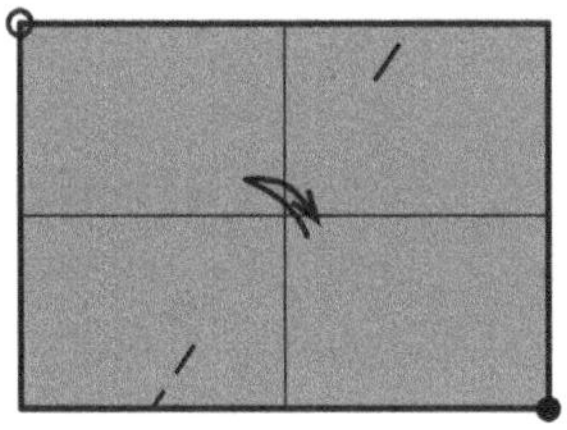

2 Bring opposite corners together and pinch the edges.

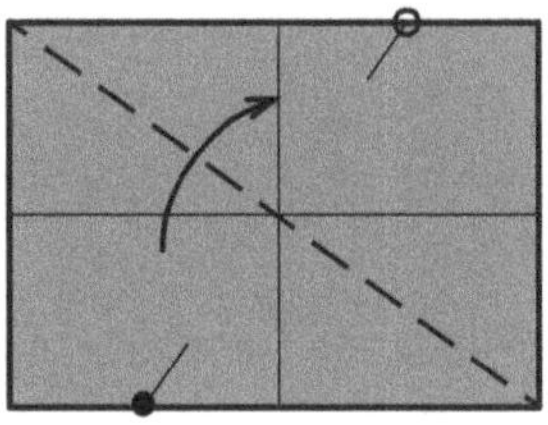

3 Fold the diagonal by joining opposite edge creases.

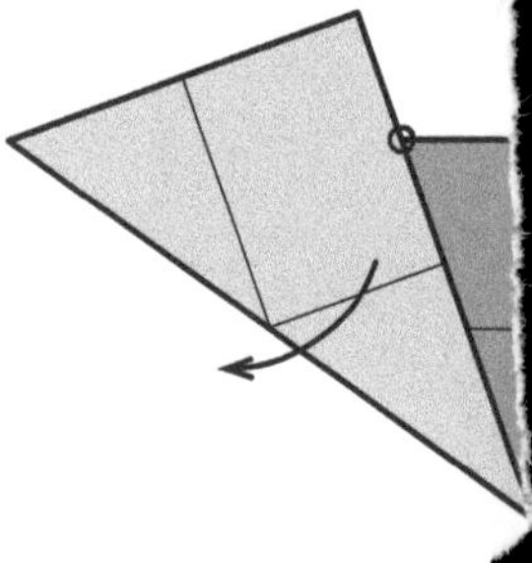

4 Unfold.

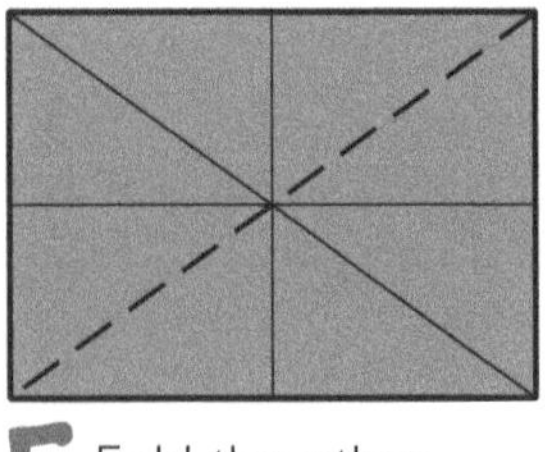

5 Fold the other diagonal using the mirror image of steps 2 to 4.

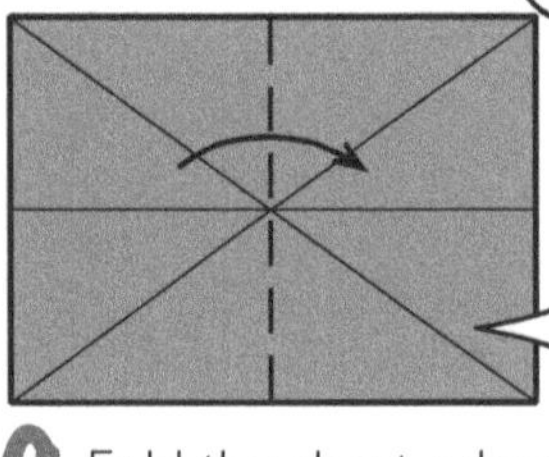

6 Fold the short edges together and rotate.

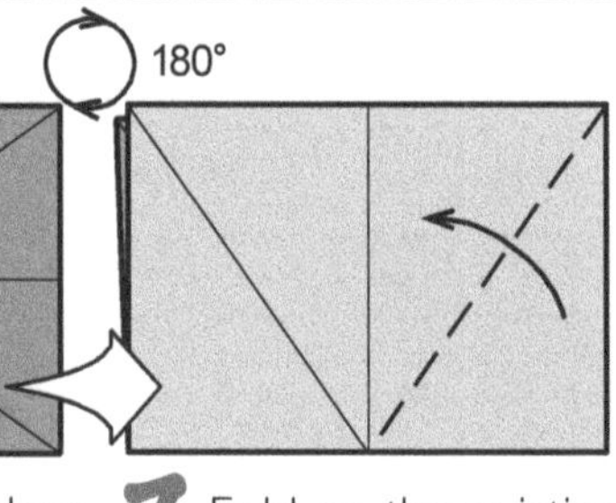

7 Fold on the existing creases.

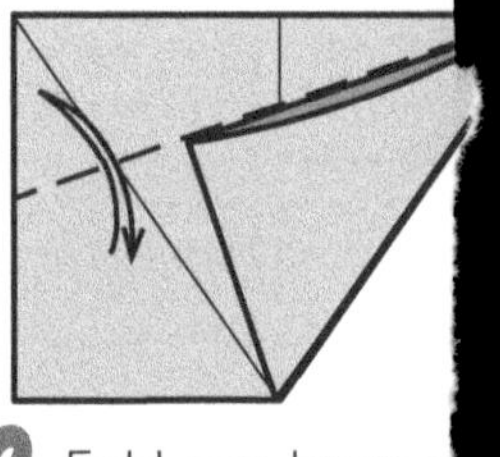

8 Fold one layer o the top edge ov the flap and unfo

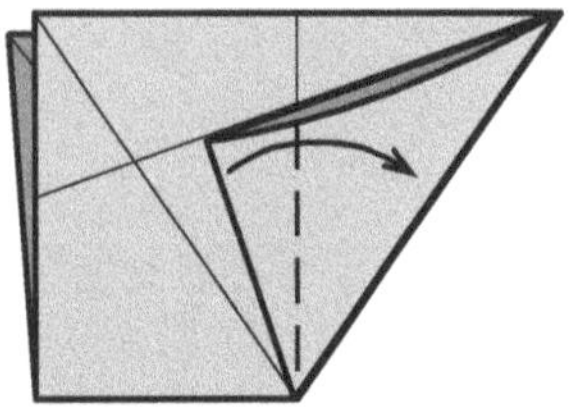

9 Fold the central flap back using the crease behind as a guide.

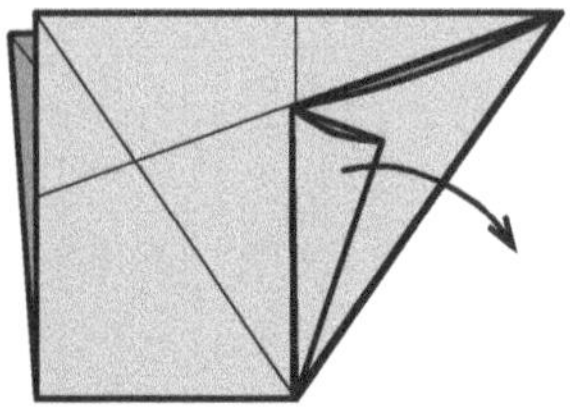

10 Unfold. Turn over and repeat steps 7 to 9.

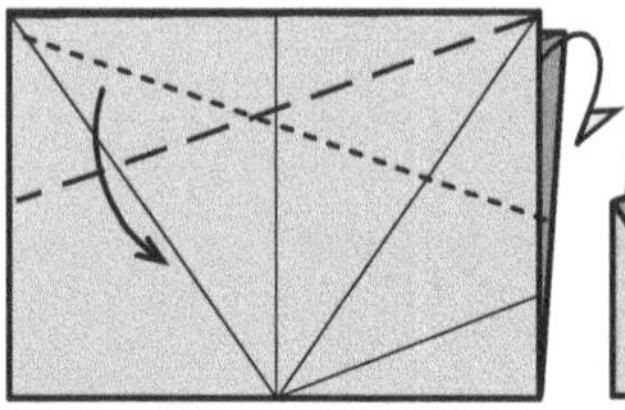

11 Fold the top flaps down on the existing creases.

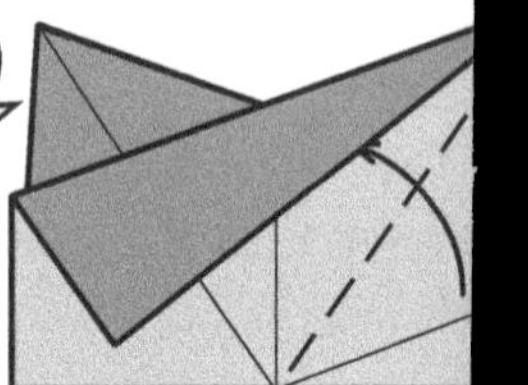

12 Tuck the right fl inside the top fl

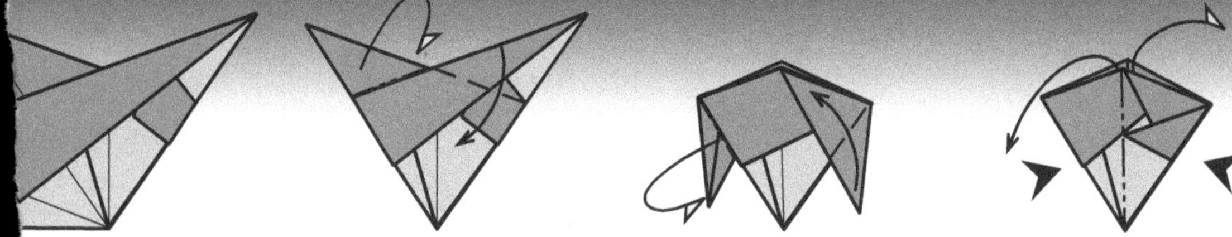

Turn over and repeat step 12.

14 Fold along the folded edge.

15 Fold along the folded edge.

16 Open out. Make 25 of these units.

lf-closed module ★★★

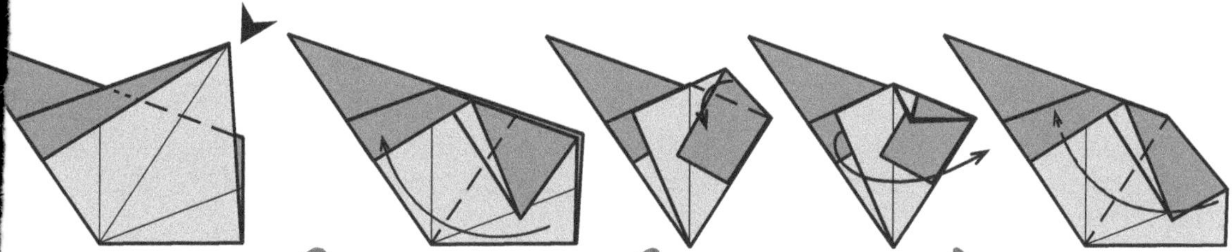

Make the module up to step 12 and turn over. Reverse fold the top right flap.

2 Fold the right flap to the left.

3 Fold the top right tip of the flap down to lie along the edge behind. Unfold the right flap.

4 Tuck the right flap inside the pocket. Follow module steps 14 to 16. Make five of these half-closed units.

ssembly ★★★★

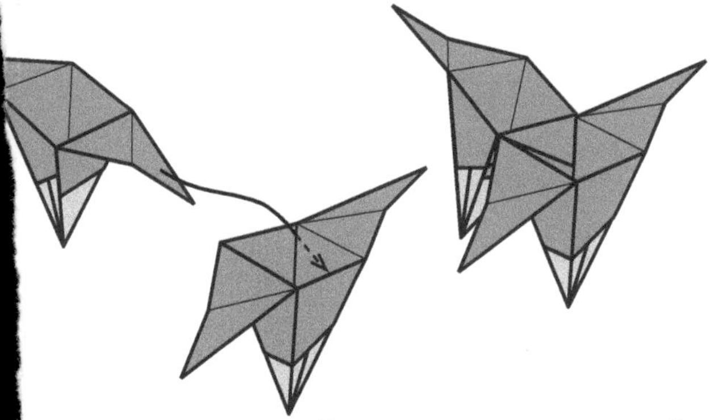

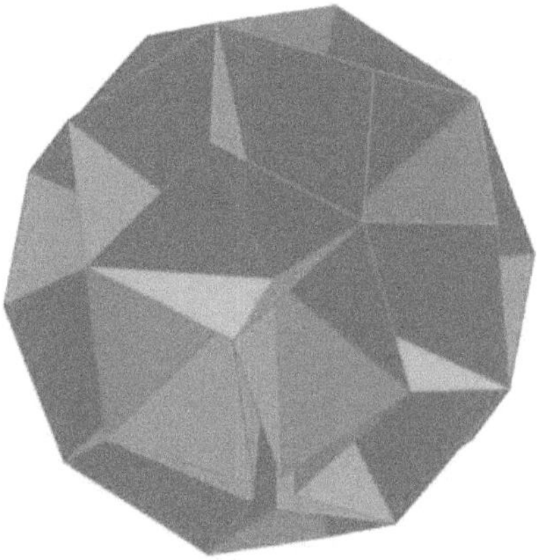

Insert the tip of the first unit into the pocket of the second unit.

2 Two units joined. The apex of each pyramid will become the centre of the model.

3 Add more units. Keep to the pattern of groups of three and five units at the vertices.

4 Finish by using the five half-closed units at final vertex.

ction

nd the vertex with the five half-closed units. Carefully spread the units and collapse e model. Do not keep the model fully flat for too long or it will lose its springy tion.

Flexicub-octahedron

★★★★☆

Use 9 square

Finished model ratio 0.

The Flexicube has been made in several origami versions. Philip Noble created one from a lo strip in the 1970s. In the early 2000s I learned Paul Slater's *Flexicube* made from train ticke Dave Brill's *Double Star Flexicube* is related to the commercially available Yoshimoto Cube Number 1. You can vary the Flexicuboctahedron design to make a truncated octahedron.

For your first attempt use 15cm origami paper: smaller paper is awkward to fold manipulate, and much larger paper makes the finished model too big for your hands to hold. Note that model needs four left-handed and four right-handed modules.

Joining Tabs ★

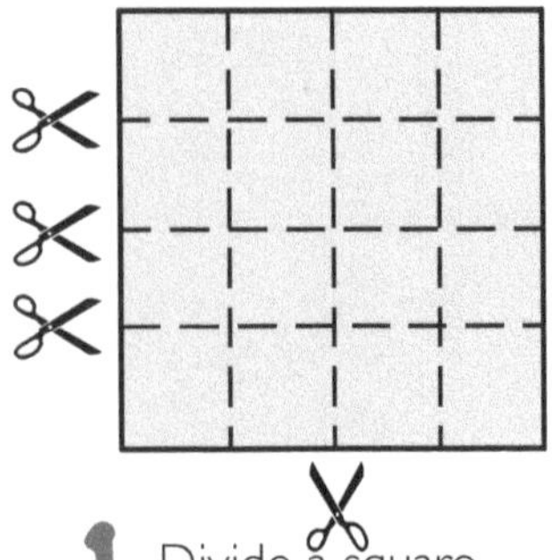

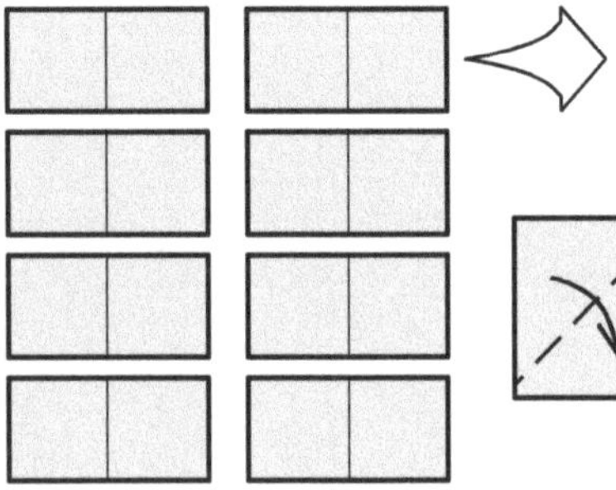

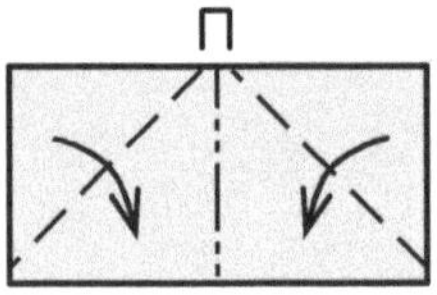

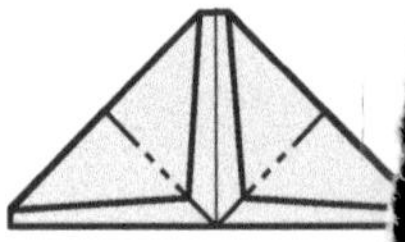

1 Divide a square into a 4x4 grid.

2 Cut into eight 2:1 rectangles.

3 Fold the corners to the centre with a 1-2 mm gap.

4 Crease to comp the joining tab.

Right-handed Module ★★★

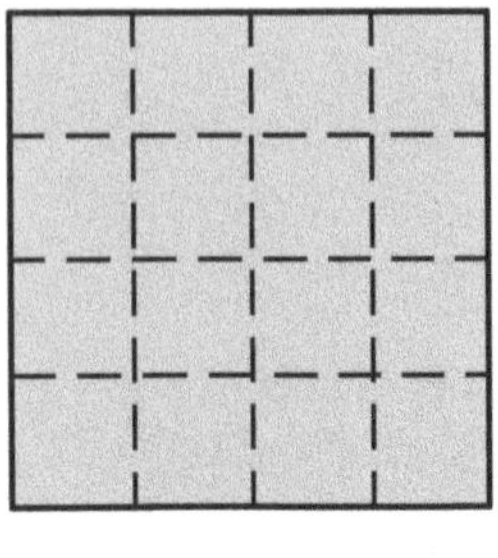

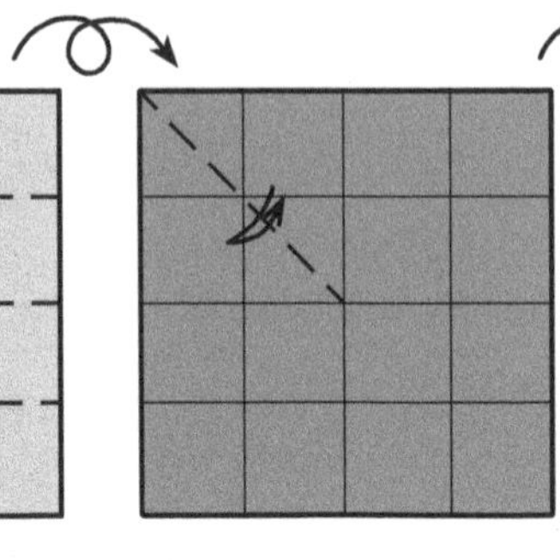

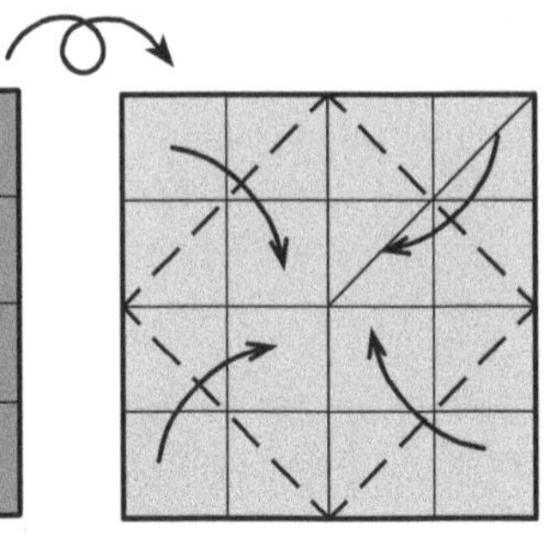

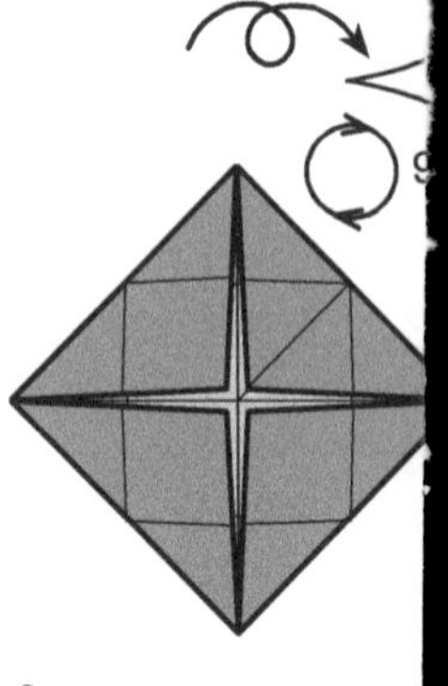

1 Divide a square into a 4x4 grid. Turn over.

2 Fold one half of a diagonal. Turn over.

3 Fold the corners to the centre.

4 Turn over.

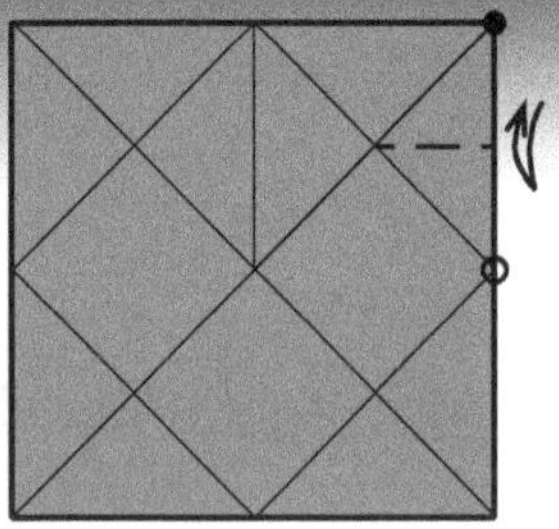

Bisect the right edge of the right triangle of the top right corner.

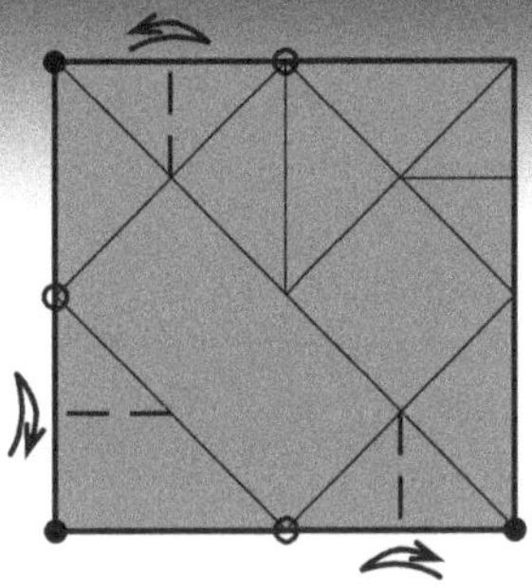

6 Repeat with rotational symmetry. Turn over and open out so that the half-diagonal runs north-east from the centre.

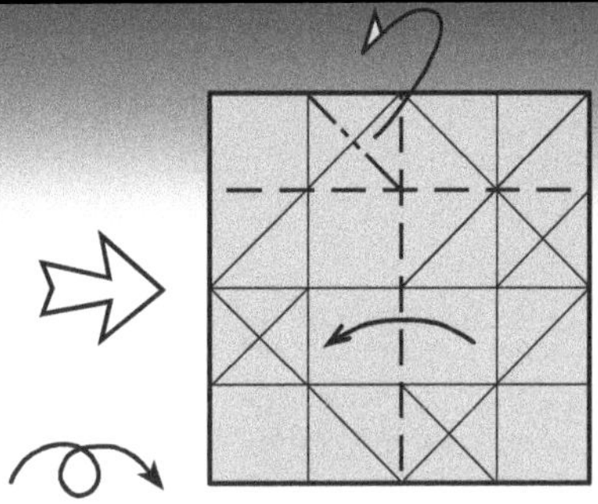

7 Fold the top mid-point behind whilst folding the paper into a box shape.

ssembly ★★★★

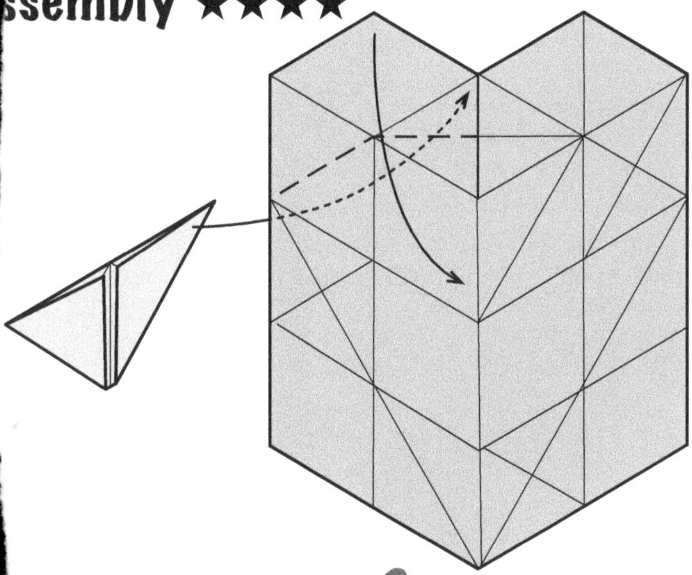

Tuck a joining tab into the pocket made in step 7.

2 Fold on the crease made step 3 to lock the tab.

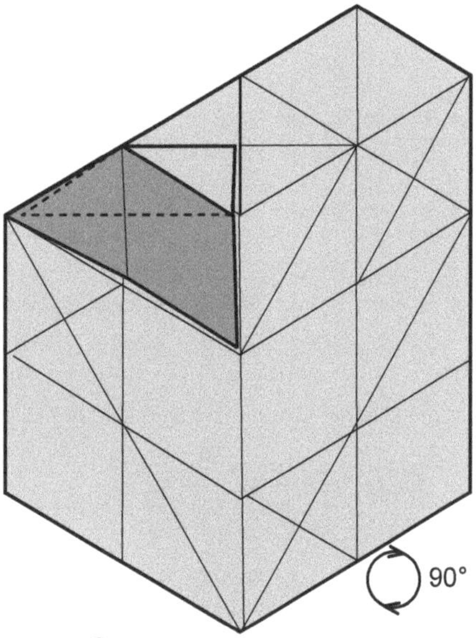

3 Rotate clockwise.

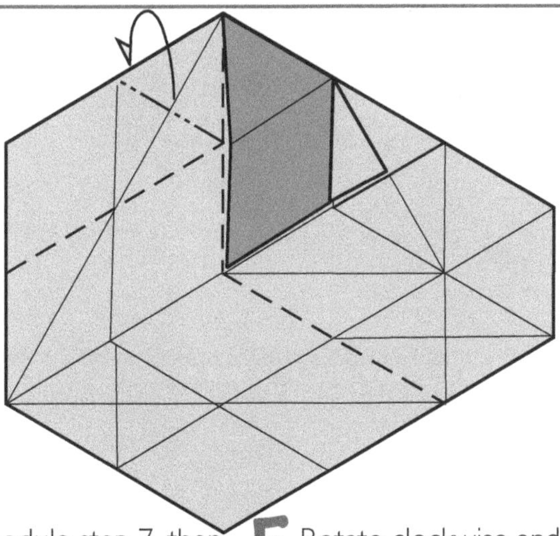

Repeat module step 7, then assembly steps 1 and 2.

5 Rotate clockwise and repeat once more.

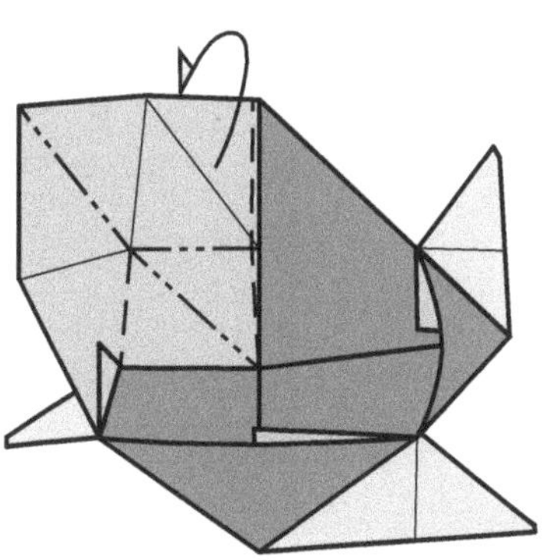

6 For the last flap, make the creases on the half-diagonal.

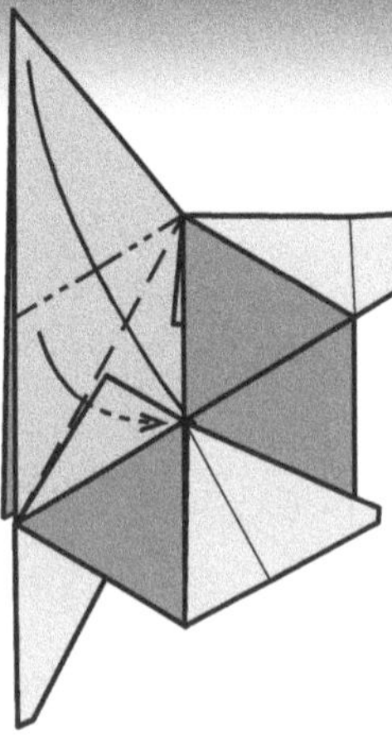

7 Fold on the crease made steps 1 and 3 to lock to tab.

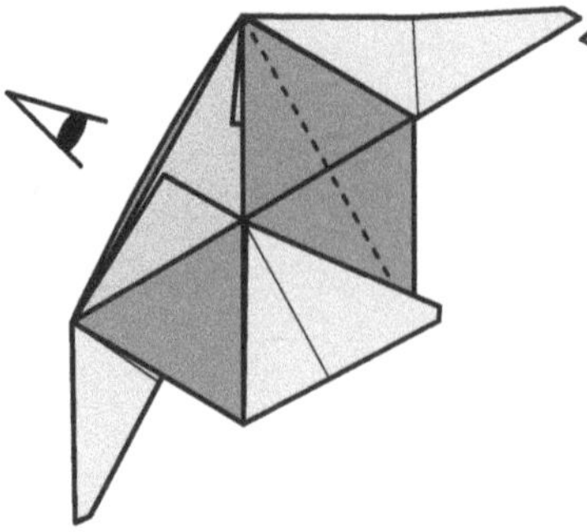

8 The next view looks at the equilateral triangle.

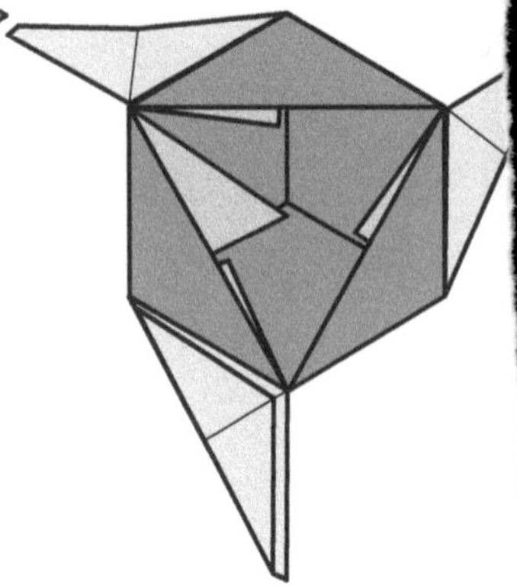

9 This right-handed module has joining tabs ready for two left-handed modules.

---

10 Make and join more units. You need four right-handed module and four left-handed modules. To make a left-handed module, follow the right-handed steps in a mirror. In particular, in steps and 6 bisect the left edge of the left triangle of the top left corner. In step 7, the half-diagonal runs north-west from the centre. Rotate the paper anticlockwise between repeats. Position and orientate the units and tabs as shown: the tab widths are exaggerated for clarity.

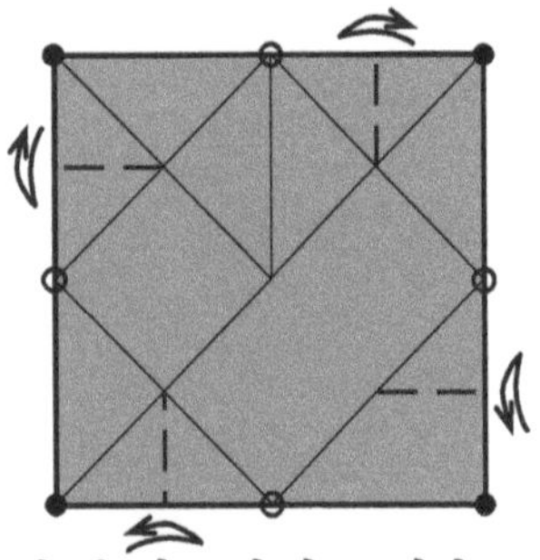

Left-handed module steps 5 and 6

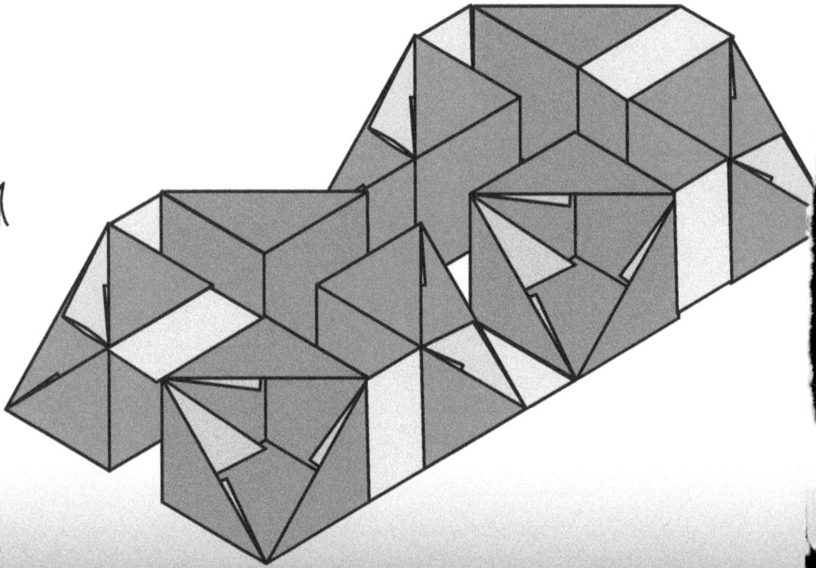

Action

Flex the model using the pairs of joining tabs. You can flex the model into several positions before repeating the cycle.

Skeletal Cube-
Rhombic Dodecahedron

★★★★☆
Use 6 sheets of A5
Finished cube edge ratio 0.5

ou need to fold these units very accurately. In particular, the crease at the centre should meet each other at one point.

odule ★★★★

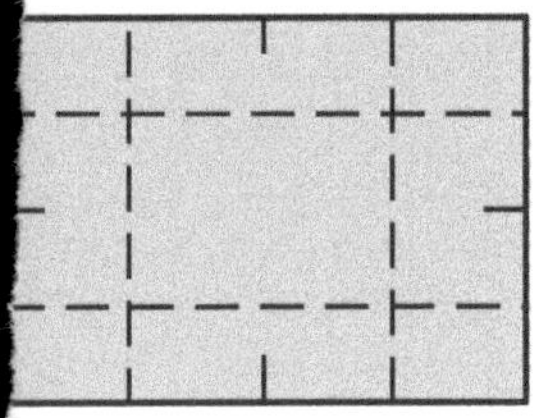

1 Pinch midpoints and then crease quarters.

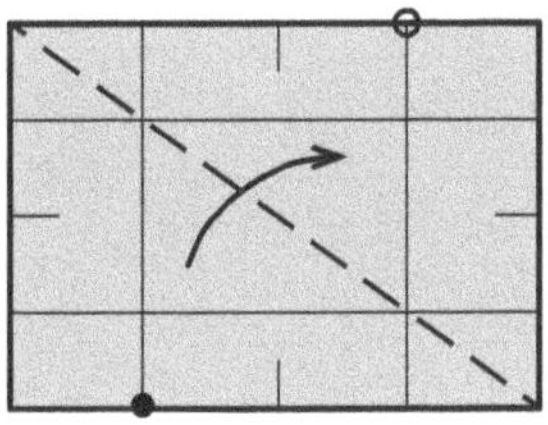

2 Fold the diagonal by joining opposite quarter marks

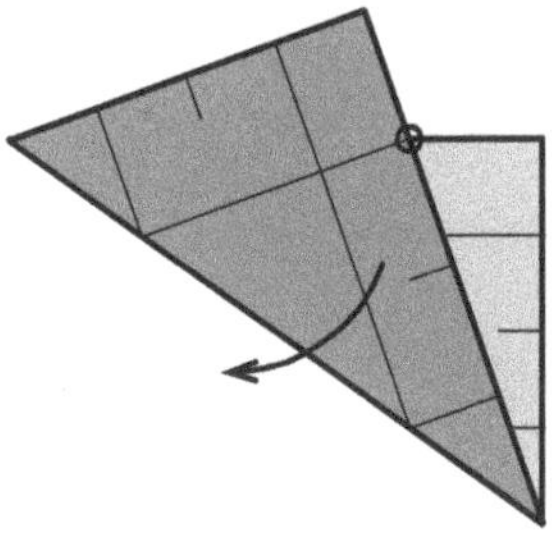

3 Unfold and repeat.

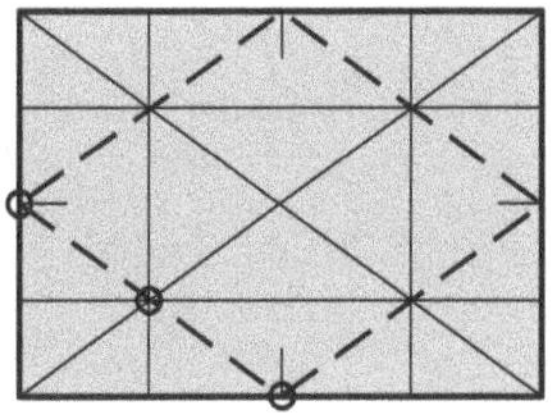

4 Fold through the midpoints and intersections of the quarter lines..

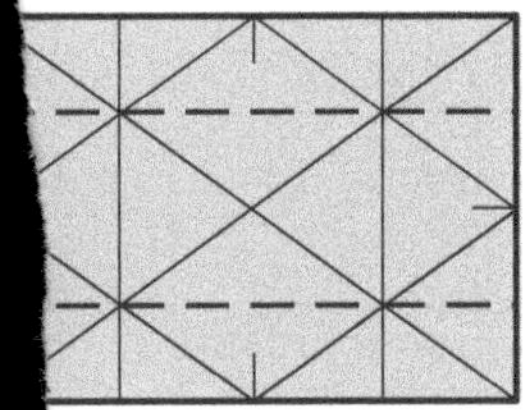

Fold the long edges to the centre.

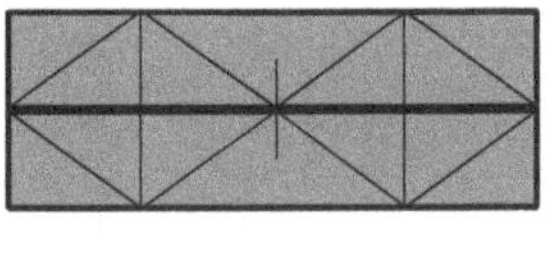

6 Turn over.

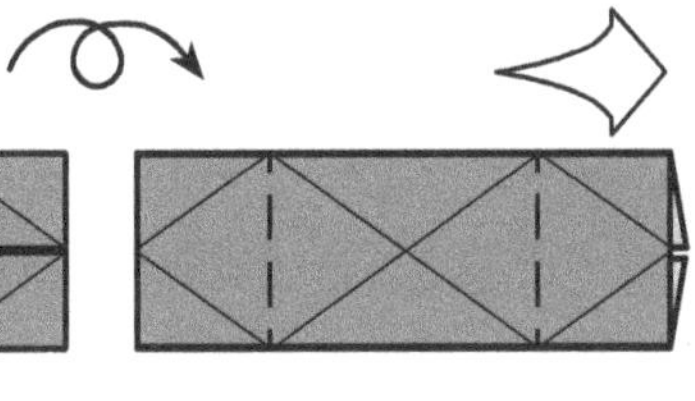

7 Fold the short edges to the centre.

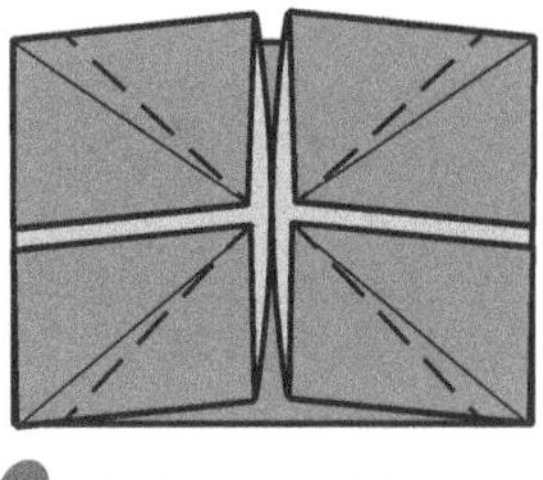

8 Fold the inside corners onto the middle line.

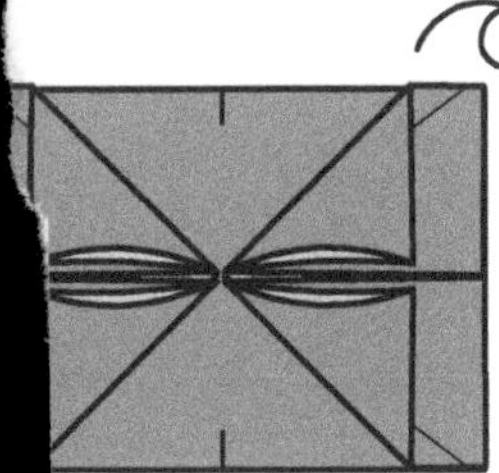

Pinch midpoints and turn over.

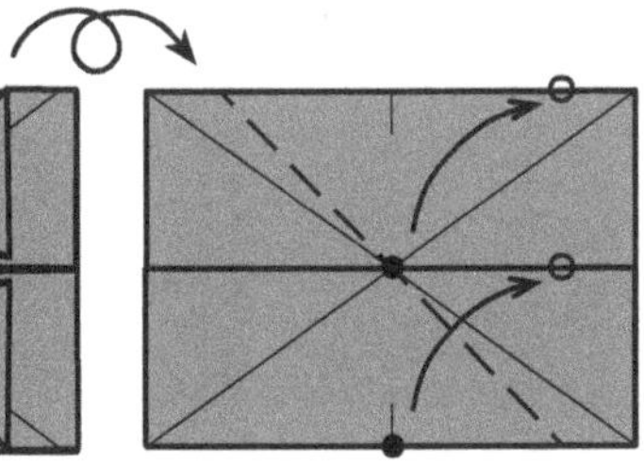

10 Fold the diagonal of the central square.

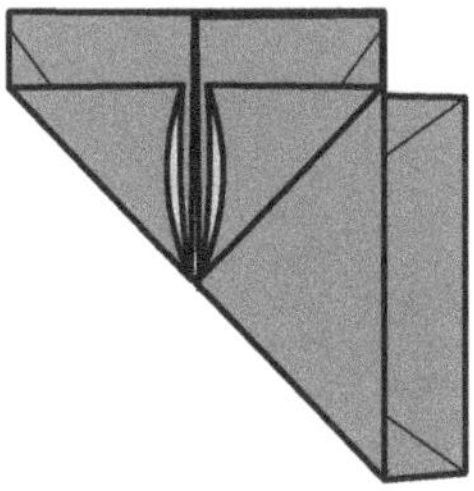

11 Unfold.

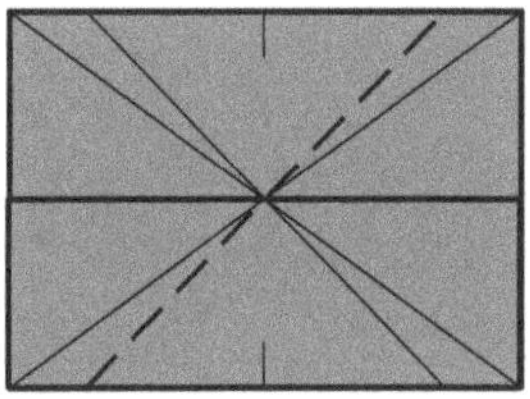

12 Repeat step 10.

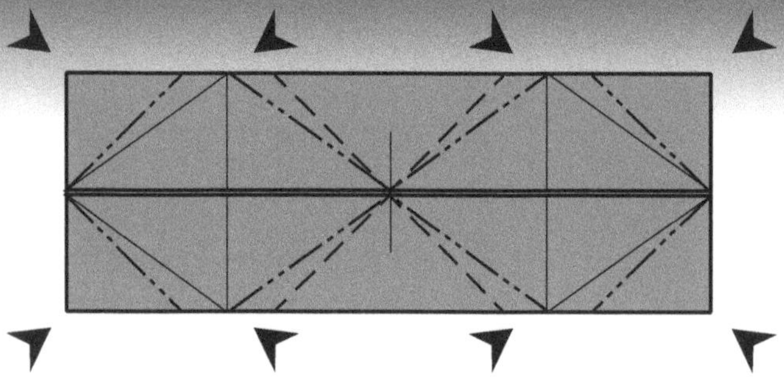

13 Sink the inside points making concave pyramid. Inside reve fold the corners.

Assembly ★★★★

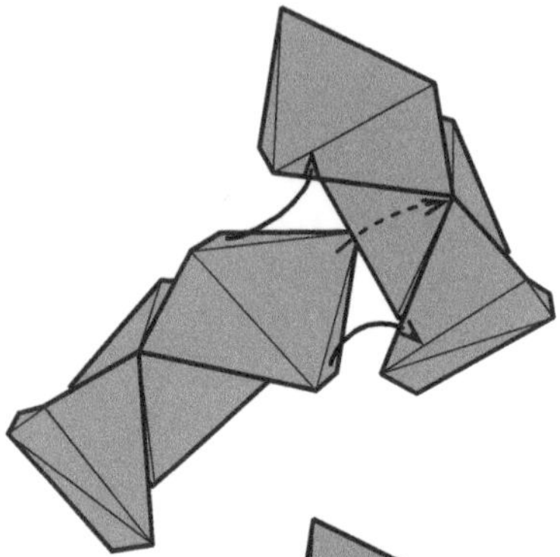

1 Insert the tip of the flap into t next unit. At the same tip tuc the two thin flaps into the tw thin side pockets of the secon unit.

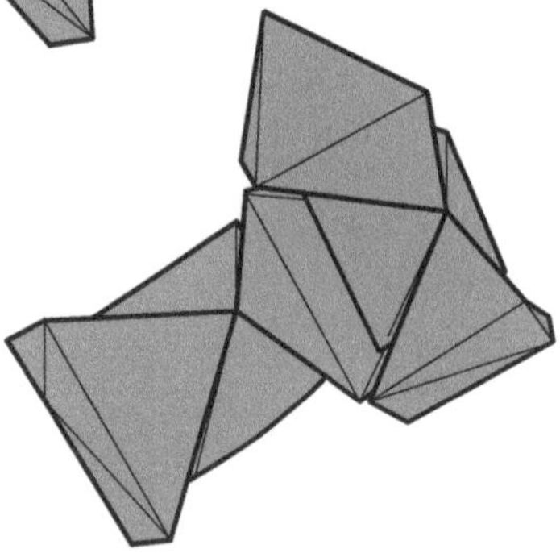

2 Two units joined. Each unit corresponds to the face of a cube, so carry on adding units the net of a cube e.g. the cros shown in the middle photo below.

Action

Flex the units so that the pyramids point in or out. The former is a skeletal cube can be made fairly secure. The latter is a rhombic dodecahedron which is not so Be careful when handling as the units can come apart, but they are easy to reatta

Cube-Regular Dodecahedron

★★★★★

Use 9 sheets of A4
Finished model ratio 1.0

his is probably the most difficult model in this book. You can make partial and soft creases on the dotted lines in steps 1, 4 and 7 to 10 to reduce unnecessary creases in the ished module. How could you make a box to hold the cube? The space inside the regular odecahedron is a cube, but what shape would fill the inside of the cube position?

lodule ★★★★★

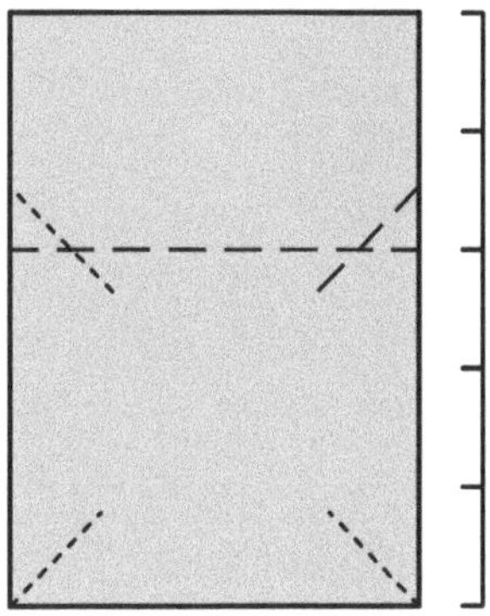

Crease three fifths up. Crease partial diagonals of the bottom square portion of the sheet.

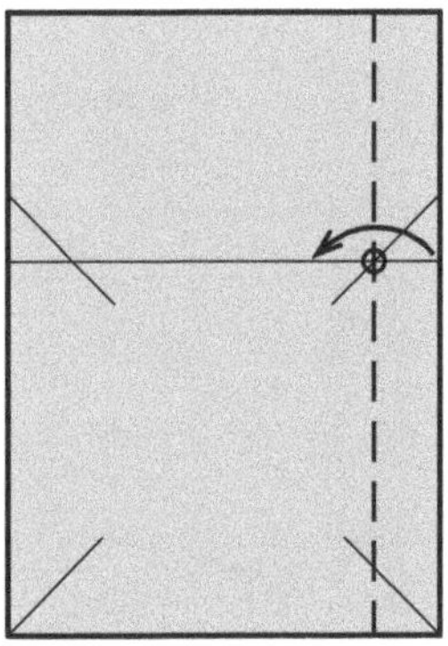

2 Fold the right edge through the intersection of the upper right diagonal and the three fifth crease.

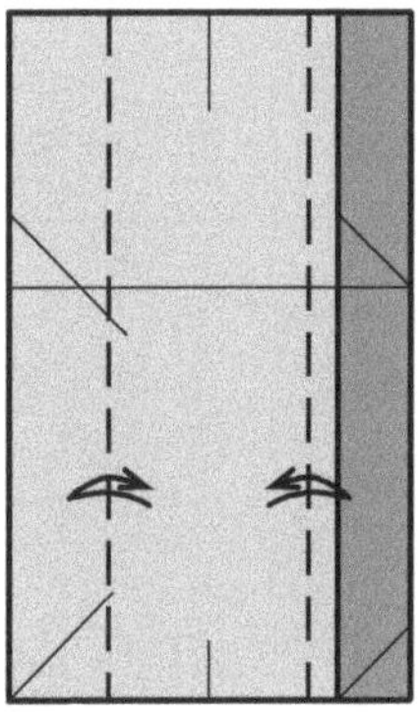

3 Pinch the midpoints of the top and bottom edges and then crease the sides in one quarter.

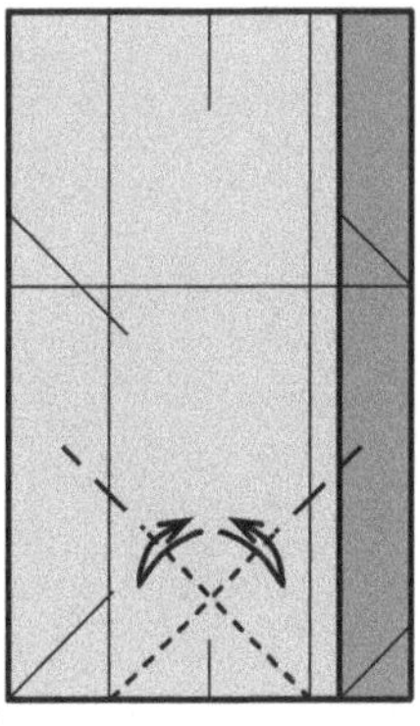

Crease the angle bisectors of the inner rectangle.

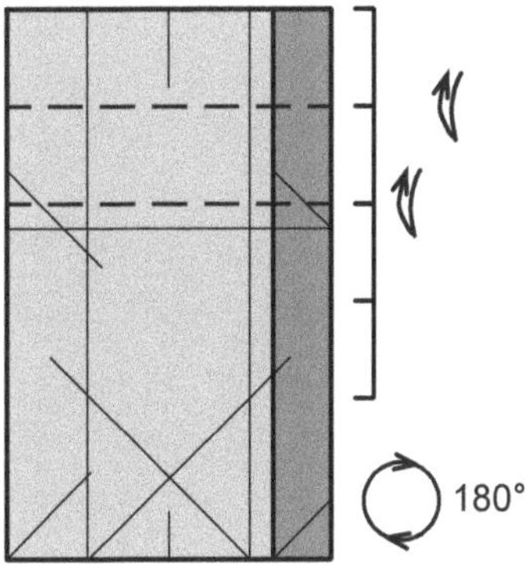

5 Use the location points to fold one half down. Unfold then fold one quarter down and unfold. Rotate 180°.

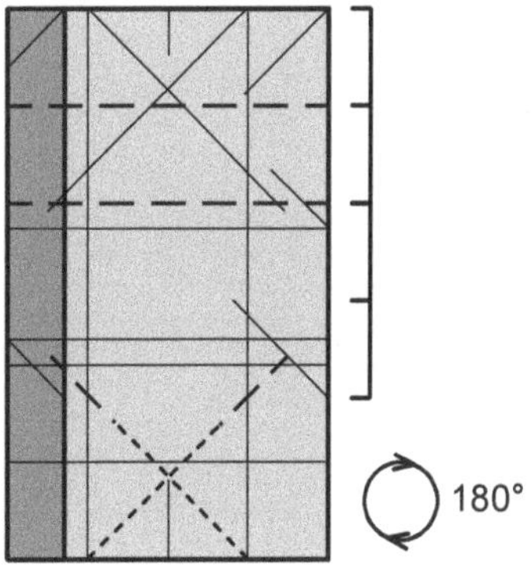

6 Repeat steps 4 and 5. Rotate 180°.

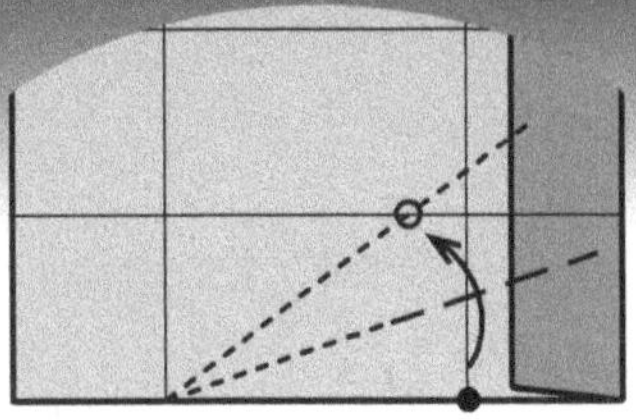

7 Fold the intersection point onto the horizontal crease.

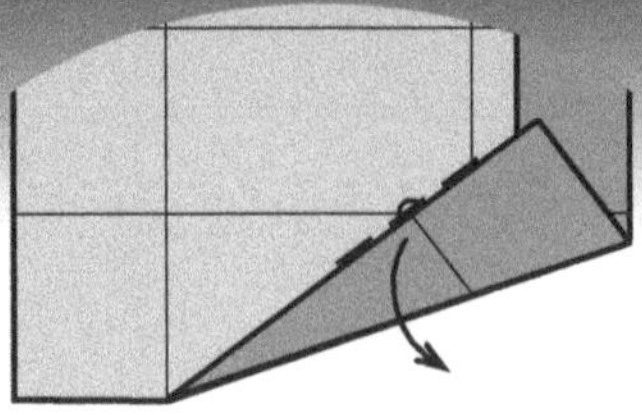

8 Mark the intersection of the raw edge and crease using your thumbnail. Unfold.

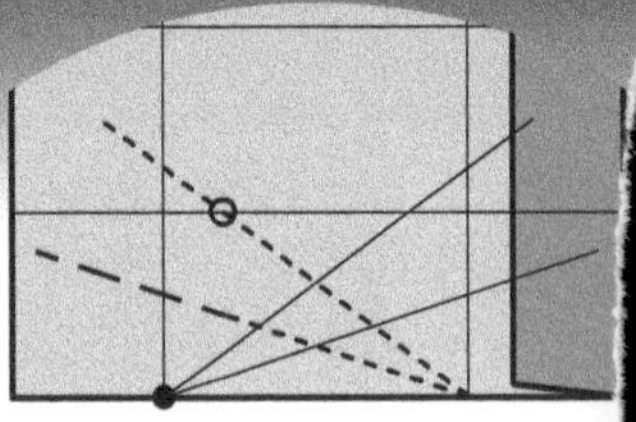

9 Repeat the mirror imag of steps 7 and 8.

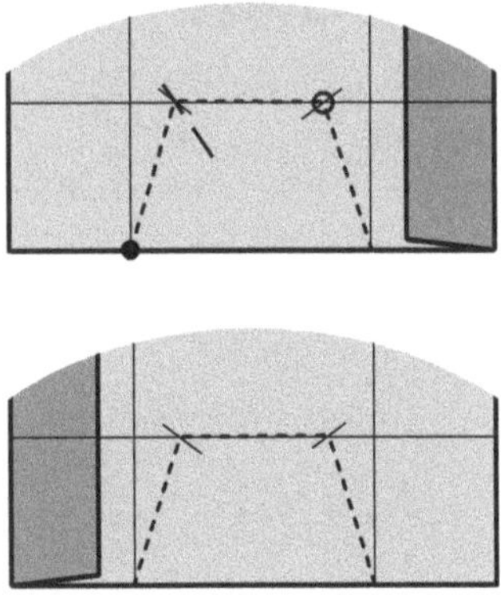

10 The three dotted lines should be the same length. Check by folding as shown. Rotate 180° and repeat steps 7 to 10.

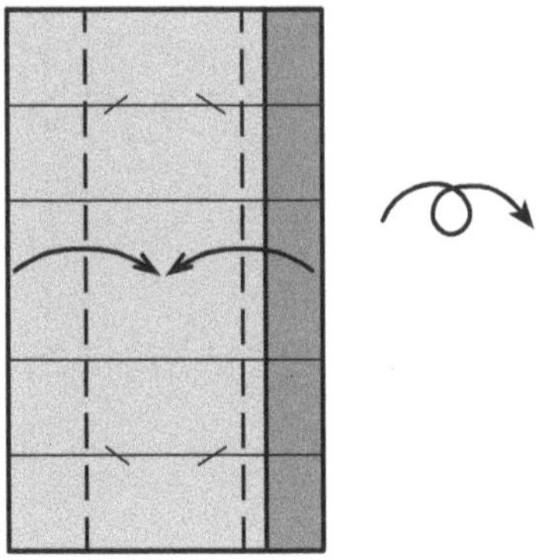

11 Fold the sides into the middle. Turn over.

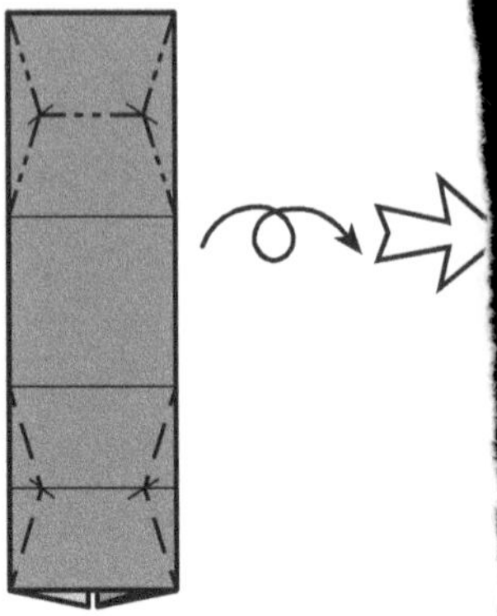

12 Crease using the location points. Unfold and turn over.

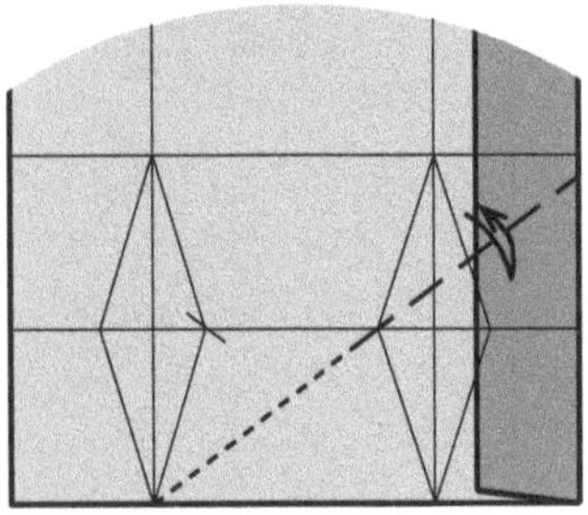

13 Extend the crease line.

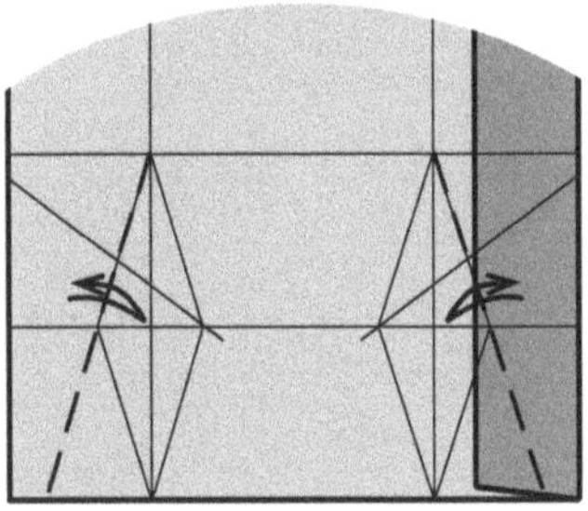

14 Repeat step 13. Extend the crease lines shown.

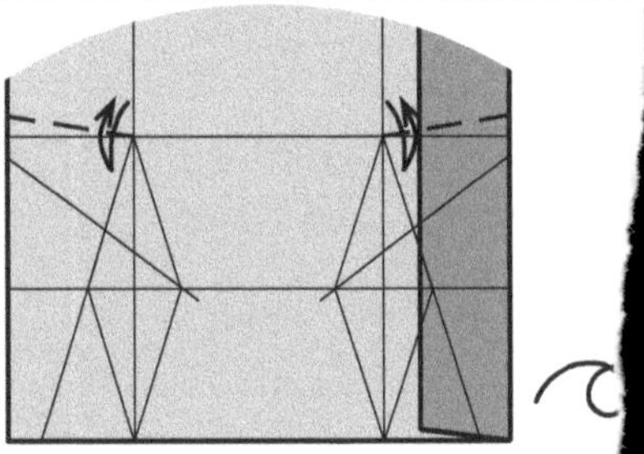

15 Crease the angle bisect Turn over.

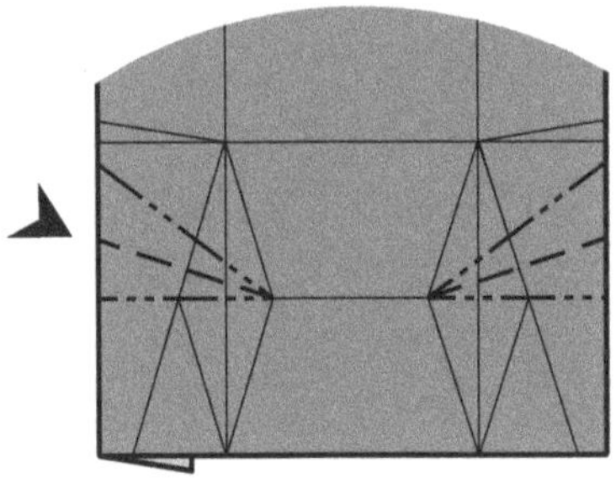

16 Reinforce the mountain folds and valley fold the angle bisectors.

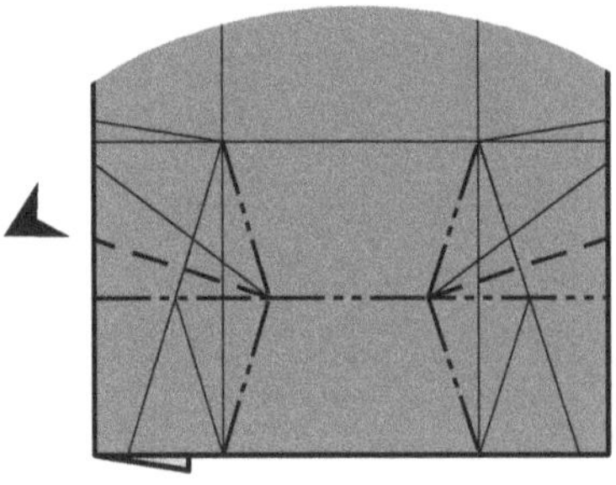

17 Crimp the folds into a 3D tent shape.

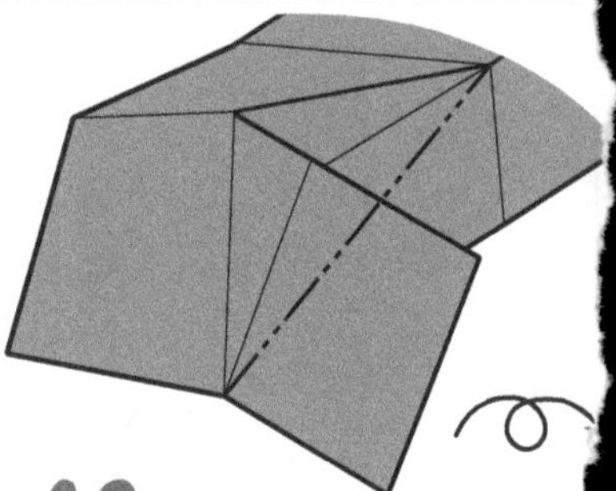

18 Mountain fold the flaps Turn over.

oining Tabs ★★

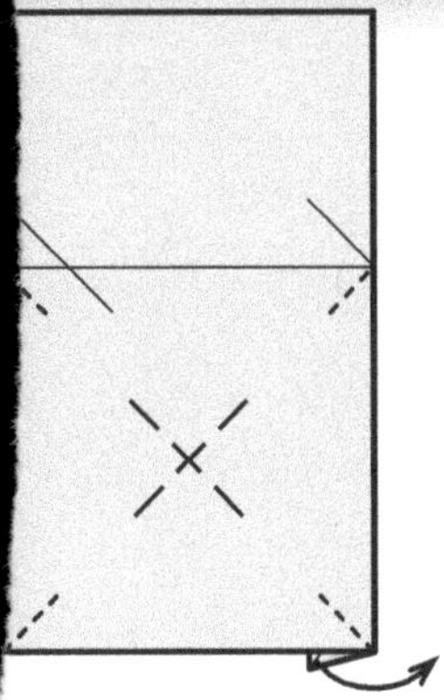

Fold step 2 of the module and turn over. Locate the centre of the lower square. Unfold the right flap.

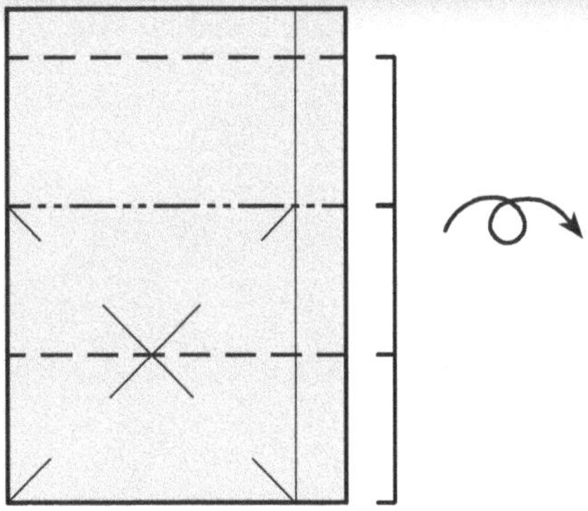

2 Pleat to obtain three equal strips. Turn over.

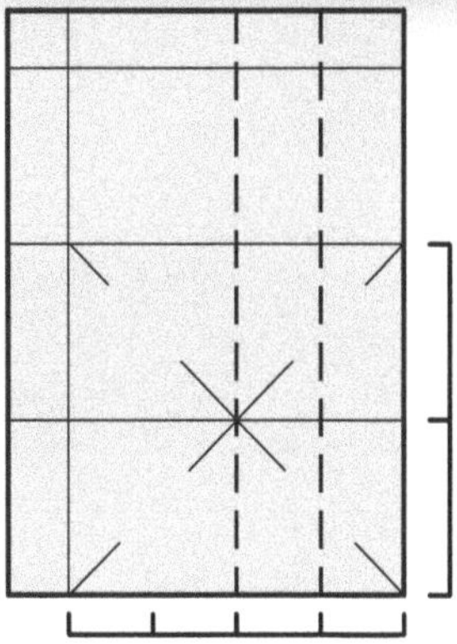

3 Fold the square in half and then quarter.

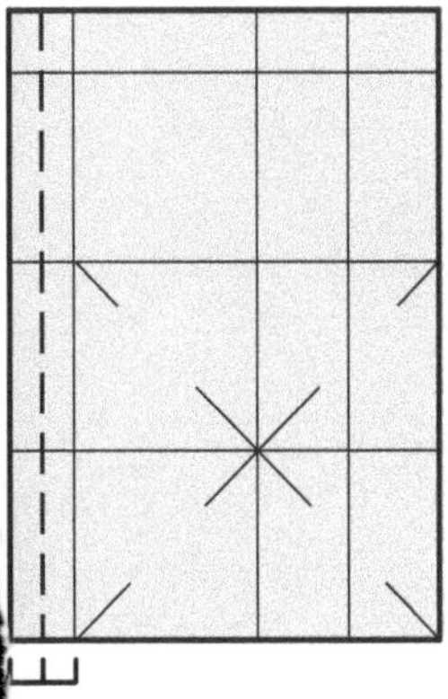

Fold the left edge to meet the first crease.

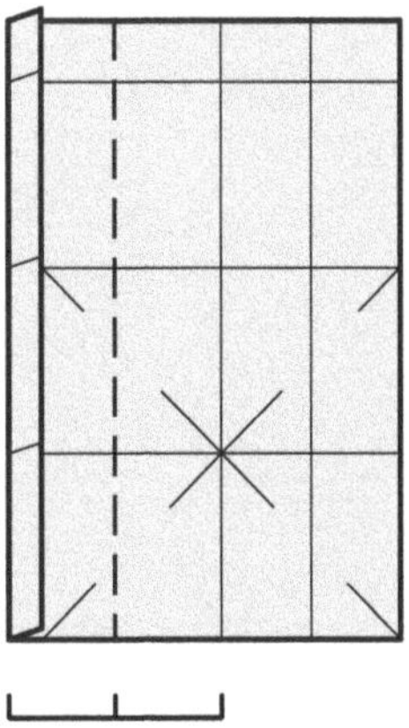

5 Fold the left edge to meet the next crease. Then unfold completely.

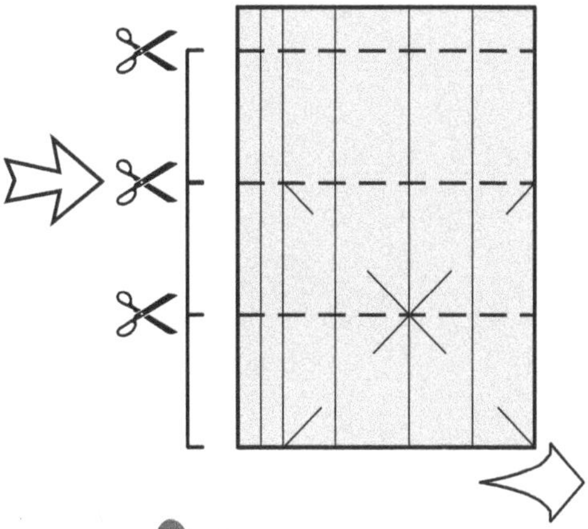

6 Cut along the horizontal creases to make three joining tabs.

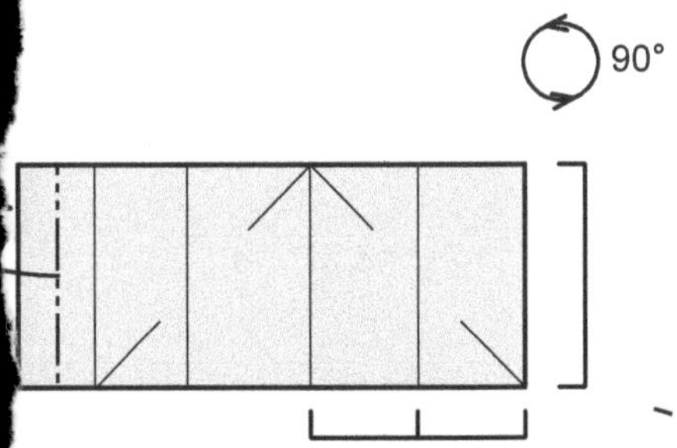

Fold the left flap behind.

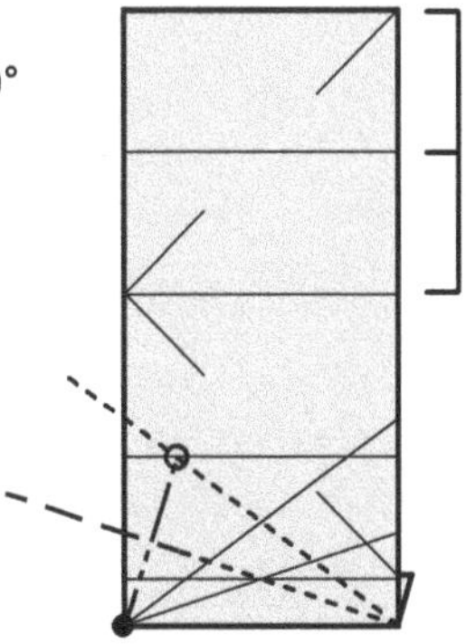

8 Use module steps 7 to 9 to make the partial regular pentagon angles.

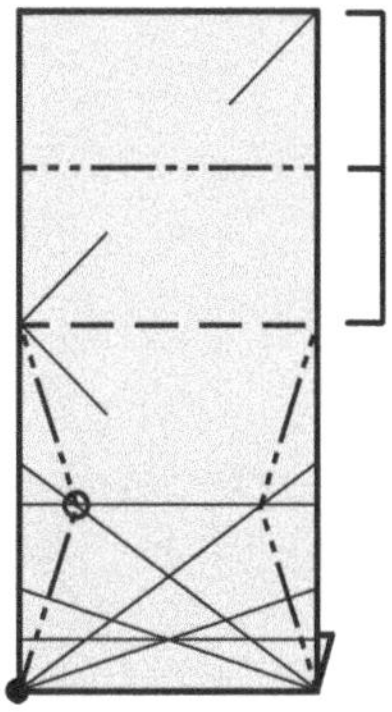

9 Final creases. You need five joining tabs in total.

Assembly ★★★★★

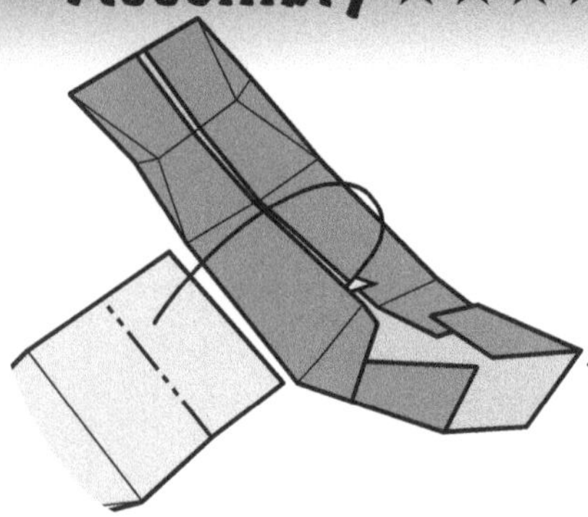

1 Mountain fold the joining tab inside the module.

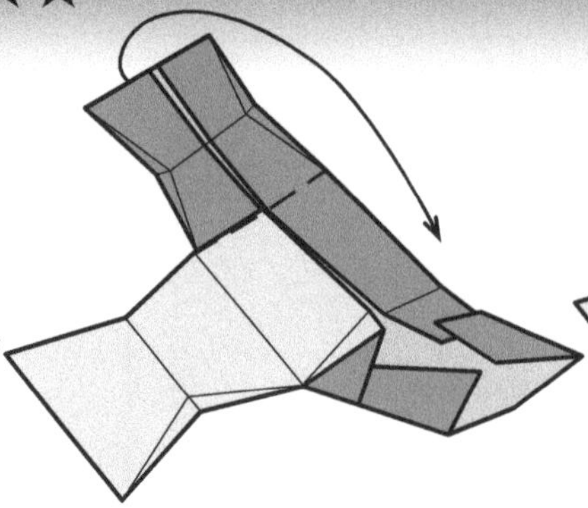

2 Fold the left side of the module over.

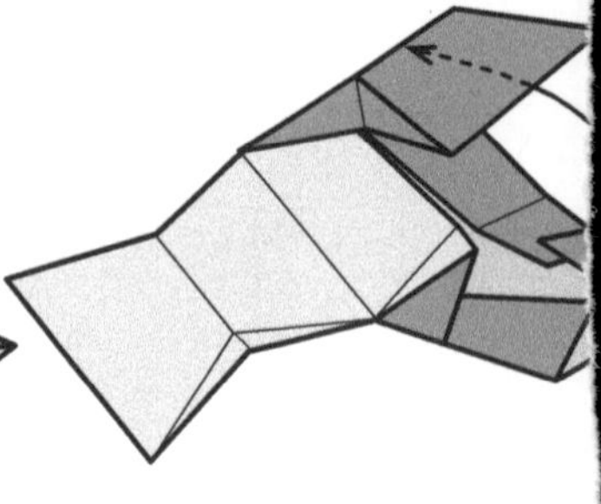

3 Tuck the left side of the module inside the right

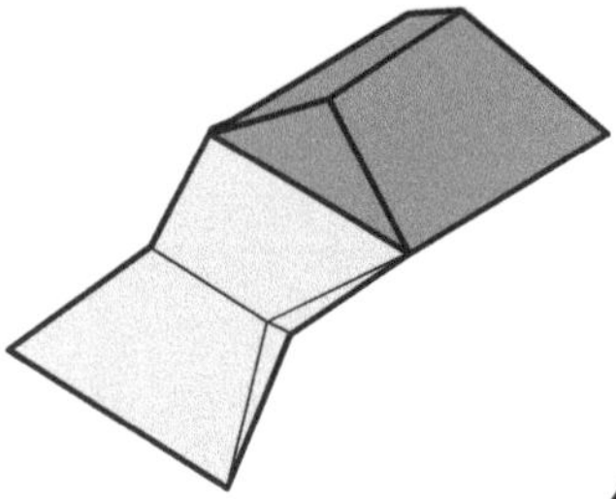

4 First unit with joining tab completed.

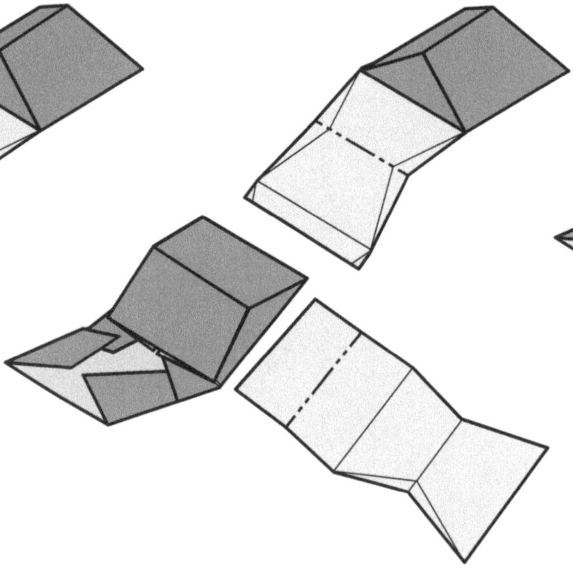

5 Make the second unit with a joining tab, then tuck the tip of first unit's joining tab into the second unit.

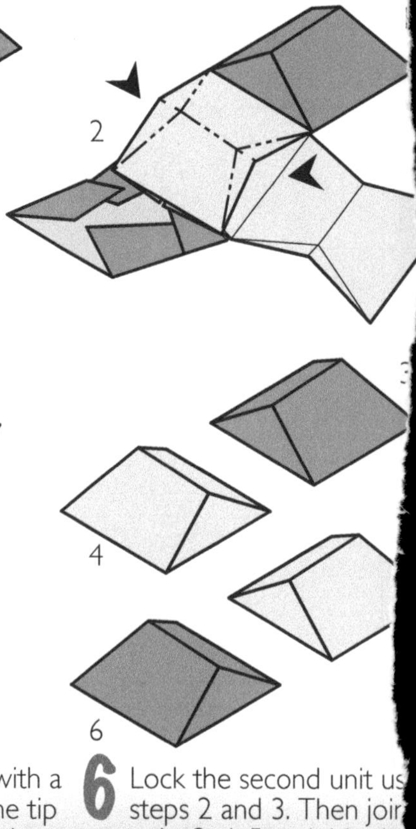

6 Lock the second unit us steps 2 and 3. Then joir units 3, 4, 5 and 6 in the order shown.

Action

Fold the units on the hinges to make the cube and regular dodecahedron.

Further Reading

I hope you enjoyed making these models, finding satisfaction, a sense of achievement and amusement (without too much frustration). I also hope the experience will stimulate your creativity and sense of wonder: there's much more action modular origami to create and explore!

Four broad categories of further reading are given here. Not all books are currently in print so you might need to look in second-hand bookshops or libraries. Try searching for websites if they become unavailable or change address. Books asterisked* are Tarquin titles available from www.tarquingroup.com.

Origami

Dave Mitchell's *Mathematical Origami** [Mitchell 97] (new edition 2018) has his *Rotating Ring of Rhombic Tetrahedra* as well many other excellent non-action modular origami. Thoki Yenn's lively diagrams give instructions for many wonderful geometric models including his *Skew Tetrahedron* [Yenn 87].

Dave Brill's *Double Star Flexicube* is in his *Brilliant Origami.* It features a wide range of intermediate to complex geometrical, animal and miscellaneous works [Brill 96].

My planar polyhedra *Octahedral Cross* and *Blintz Icosidodecahedron* were compiled with other planar modular models by Dave Petty in a publication by the British Origami Society [Petty 13]. *Intersecting Planes Gallery* shows *WXYZ* and other planar models [Mukerji 03].

Robert Lang collected diverse action origami [Lang 97]. None of the models are purely geometric, however. Jeremy Shafer diagrams and gives performance tips for his action origami [Shafer 12]. Yami Yamauchi's website has his *Fireworks, Jacob's ladder* and *Doodlebug* [Michaelis 14].

Kunihiko Kasahara shows the skeletal octahedron as a small whirling top in *Origami Omnibus* [Kasahara 88]. Kasahara's *Panorama Cube* is in the same book.

Good books of modular origami include those by Tomoko Fuse (action [Fuse 87] and non-action [Fuse 91]) and Miyuki Kawamura [Kawamura 02].

Mathematics and Geometry

Two valuable books that have been a source of inspiration for many years are David Wells' *Curious and Interesting Geometry* [Wells 91] which gathers diverse geometrical theorems, oddities and facts, and Cundy and Rollett's classic *Mathematical Models* which describes, analyses and organises the regular and semi-regular polyhedra [Cundy and Rollett 81].

Other books include a guide of instructions for making many complicated polyhedra [Wenninger 71], an accessible summary of polyhedra, their properties and the relationships between them [Cromwell 04] and a book aimed at the layperson that includes many intriguing pictures and their associated mathematics [Steinhaus 00].

Books of action models to cut out and make include *Metamorphs** by Robert Byrnes [Byrnes 99] and *The Magic of Flexagons** by Dave Mitchell [Mitchell 98]. The former has 3D as well as 2D models that work by folding.

The Jitterbug is in *A Fuller Explanation* [Edmondson 87], an accessible introduction to Fuller's geometric and mathematical work.

The late and great Martin Gardner is well known for his accessible works of recreational mathematics. Almost any are worth reading, but keep an eye open for those that reproduce his original Scientific American article on hexaflexagons [Gardner 91].

Origami and Mathematics

This field has grown quickly in the last few years but has much older roots: one book was originally written in 1893 [Sundara Row 66]. The mathematics of dividing lengths into equal parts is accessible to those with mathematics at secondary school level [Hatori] [Guy 75].

More advanced works include a book on how to design certain kinds of complex single sheet origami [Lang 03], a research monograph [Demaine and O'Rourke 07] and the proceedings of the International Meetings of Origami Science, Mathematics, and Education (OSME)* [Lang 09] [Wang-Iverson et al. 11].

Origami and Mathematics Education

Several books use origami for teaching and learning mathematics. These can be divided by level (school or college/university) and type (descriptions of activities or descriptions with specific pedagogic advice).

Project Origami has worksheets for undergraduates with pedagogical notes for teachers but several activities could be adapted for secondary schools [Hull 06]. The OSME proceedings usually have several relevant papers [Lang 09] [Wang-Iverson et al. 11].

School-level books are typically at secondary school level (11 to 16 years of age). Two publications of activities by the National Council of Teachers of Mathematics have largely the same selection of classic material [Olson 75] [Johnson 99]. *Geometric Exercises in Paper Folding* is a little dated and hard to understand but is of historical interest [Sundara Row 66]. Kazuo Haga presents several elegant investigations in *Origamics* [Haga et al. 08].

Learning Mathematics with Origami is aimed at primary and secondary school teachers [Lam and Pope 16]. As well as giving practical advice on using origami for learning mathematics, it promotes opportunities for learners to think creatively, justify decisions, deepen understanding, take responsibility for their own learning and work collaboratively with others.

Bibliography

[Brill 96] David Brill. *Brilliant Origami: A Collection of Original Designs*. Japan Publications, 1996. 77

[Byrnes 99] Robert Byrnes. *Metamorphs: Transforming Mathematical Surprises*. Tarquin Publications, 1999. 78

[Cromwell 04] Peter Cromwell. *Shapes in Space: Convex Polyhedra with Regular Faces*. Association of Teachers of Mathematics, 2004. 78

[Cundy and Rollett 81] H. Martyn Cundy and A. R. Rollett. *Mathematical Models*, Third edition. Tarquin Publications, 1981. 77

[Demaine and O'Rourke 07] Erik D Demaine and Joseph O'Rourke. *Geometric Folding Algorithms: Linkages, Origami, Polyhedra*. Cambridge University Press, 2007. 78

[Edmondson 87] Amy C. Edmondson. *A Fuller Explanation: the Synergetic Geometry of R. Buckminster Fuller*. Design science collection, Birkhäuser, 1987. 78

[Fuse 87] Tomoko Fuse. *Let's play*. Tankobon, 1987. In Japanese. 77

[Fuse 91] Tomoko Fuse. *Unit Origami: Multidimensional Transformations*. Japan Publications, 1991. 77

[Gardner 91] Martin Gardner. *Mathematical Puzzles and Diversions*. Penguin Books Ltd, 1991. 78

[Guy 75] Mick Guy. *Geometrical Division*. British Origami Society, 1975. 78, 88

[Haga et al. 08] Kazuo Haga, Josefina Fonacier, and Masami Isoda. *Origamics: Mathematical Explorations Through Paper Folding*. World Scientific, 2008. 78

[Hatori] Koshiro Hatori. 'How to Divide the Side of Square Paper.' Available online (`http://www.origami.gr.jp/Archives/People/CAGE_/divide/index-e.html`). 78, 88

[Hull 06] Thomas Hull. *Project Origami: Activities for Exploring Mathematics*. A.K. Peters, 2006. 78

[Johnson 99] Donovan A Johnson. *Paper Folding for the Mathematics Class*. National Council of Teachers of Mathematics, 1999. 78

[Kasahara 88] Kunihiko Kasahara. *Origami Omnibus: Paper-Folding for Everybody*. Japan Publications, 1988. 77

[Kawamura 02] Miyuki Kawamura. *Polyhedron Origami for Beginners.* Japan Publications Trading Co, 2002. 77

[Lam and Pope 16] Tung Ken Lam and Sue Pope. *Learning Mathematics with Origami.* Association of Teachers of Mathematics, 2016. 78

[Lang 97] Robert J. Lang. *Orgami in Action: Paper Toys That Fly, Flap, Gobble, and Inflate!*, St Martin's Griffin edition edition. St Martin's Press, 1997. 77

[Lang 03] Robert J. Lang. *Origami Design Secrets: Mathematical Methods for an Ancient Art.* AK Peters, 2003. 78

[Lang 09] Robert J Lang, editor. *Origami 4: Fourth International Meeting of Origami Science, Mathematics, and Education.* A K Peters, 2009. 78

[Michaelis 14] Mari J Michaelis. 'Yami Yamauchi - Yami's Corner.' Available online (`http://www.spitenet.com/origami/yami.shtml`), 2014. 77

[Mitchell 97] David Mitchell. *Mathematical Origami: Geometrical Shapes by Paper Folding.* Tarquin, 1997. 77

[Mitchell 98] David Mitchell. *The Magic of Flexagons: Manipulative Paper Puzzles to Cut Out and Make.* Tarquin Publications, 1998. 78

[Mukerji 03] Meenakshi Mukerji. 'Intersecting Planes Gallery.' Available online (`http://www.origamee.net/gallery/planars.html`), 2003. 77

[Olson 75] A. T Olson. *Mathematics Through Paper Folding.* National Council of Teachers of Mathematics, 1975. 78

[Petty 13] David Petty. *Planar Modular Origami.* British Origami Society, 2013. 77

[Shafer 12] Jeremy Shafer. *Origami Ooh La La! Action Origami for Performance and Play.* CreateSpace Independent Publishing Platform, 2012. 77

[Steinhaus 00] H. Steinhaus. *Mathematical Snapshots*, Third edition. Dover Publications Inc., 2000. 78

[Sundara Row 66] T. Sundara Row. *Geometric Exercises in Paper Folding.* Dover Publications, 1966. 78

[Wang-Iverson et al. 11] Patsy Wang-Iverson, Robert J. Lang, and Mark Yim, editors. *Origami 5: Fifth International Meeting of Origami Science, Mathematics, and Education*, First edition. A K Peters/CRC Press, 2011. 78

[Wells 91] David G. Wells. *The Penguin Dictionary of Curious and Interesting Geometry.* Penguin, 1991. 77

[Wenninger 71] Magnus J. Wenninger. *Polyhedron Models.* Cambridge University Press, 1971. 78

[Yenn 87] Thoki Yenn. *Orikata.* British Origami Society, 1987. 77

Index

1

12-piece Carousel, 22

3

3-piece Square Spinner, 41
3-piece Trihexaflexagon, 15
3D Magic Star, 52

4

4-piece Harlequin Slider, 4

B

Beak Unit, 48

C

Collapsoid, 66
x Dodecahedron, 73

D

Double Skew Tetrahedra, 26

F

Flexicuboctahedron, 68
Fluxicube: Rotating Ring of Six Cubes, 39
Four Cube Flexagon, 36
Four Square Flexagon, 35

H

Hexagonal Slider, 10

J

Jacob's Ladder, 33
Jitterbug, 58

M

Magic Star: spiral version, 1
Magic Wallet, 29
Multiform Flipper, 24

O

Octa Slider 3D Wheel, 54
Octagon Waterwheel, 50

P

Petrie-Coxeter Honeycomb, 61

R

Radioactive Ball, 55
Ribbon Slider, 12
Rotating Octagram Ring, 20

S

Shaky Shuriken Slider, 6
Six Cube Flexagon (Pseudo-Fluxicube), 38
Skeletal Cube-Rhombic Dodecahedron, 71
Skeletal Octahedron, 44
Square Slider, 8

W

Wobbling Wall, 40
WXYZ, 46

Milton Keynes UK
Ingram Content Group UK Ltd.
UKHW022319280624
444841UK00004B/28

9 781911 09394